Funding and Financing Transport Infrastructure

T0173909

This book seeks to enhance understanding of the impacts of project setup and its implementation environment on project performance by leveraging information from the study of a rich set of European transport infrastructure project cases. It puts forward a system's view of project delivery and aims to serve as a strategic tool for decision makers and practitioners. The proposed approach is not limited to specific stakeholder views. On the contrary, it allows stakeholders to formulate their own strategies based on an holistic set of potential implementation scenarios.

Furthermore, by including cases of projects that have been influenced by the recent financial crisis, the book aims to capitalise on experiences and provide guidelines as to the design and implementation of resilient projects delivered both through traditional as well as Public Private Partnership (PPP) models.

Finally, the book proposes a new Transport Infrastructure Resilience Indicator and a corresponding project rating system that can be assessed with an eye to the future, ultimately aiming to support the successful delivery of transport infrastructure projects for all stakeholders involved.

Athena Roumboutsos is an Associate Professor in the Department of Shipping, Trade and Transport at the University of the Aegean, Greece.

Hans Voordijk is Associate Professor in Supply Chain Management at the department of Construction Management and Engineering and director of the PDEng program in Civil Engineering at Twente University, the Netherlands.

Aristeidis Pantelias is a Senior Lecturer at the Bartlett School of Construction and Project Management and Programme Leader, MSc Infrastructure Investment and Finance, University College London, Faculty of the Built Environment, UK.

Spon Research

Publishes a stream of advanced books for built environment researchers and professionals from one of the world's leading publishers. The ISSN for the Spon Research programme is ISSN 1940–7653 and the ISSN for the Spon Research E-book programme is ISSN 1940–8005

Funding and Financing Transport Infrastructure
978-1-138-29389-2
A. Roumboutsos, H. Voordijk and A. Pantelias

Valuing People in Construction
978-1-138-20821-6
F. Emuze and J. Smallwood

Target Cost Contracting Strategy in Construction: Principles, Practices and Case Studies
978-1-315-62453-2
D.W.M Chan & J.H.L. Chan

New Forms of Procurement: PPP and Relational Contracting in the 21st Century
978-1-138-79612-6
M. Jefferies & S. Rowlinson

Trust in Construction Projects
978-1-138-81416-5
A. Ceric

Fall Prevention Through Design in Construction: The Benefits of Mobile Construction
978-1-138-77915-0
I. Kamardeen

The Soft Power of Construction Contracting Organisations
978-1-138-80528-6
S.O. Cheung, P.S.P. Wong & T.W. Yiu

Leadership and Sustainability in the Built Environment
978-1-138-77842-9
A. Opoku & V. Ahmed

FRP-Strengthened Metallic Structures
978-0-415-46821-3
X. L. Zhao

Funding and Financing Transport Infrastructure

Business Models to Enhance and Enable Financing of Infrastructure in Transport

Edited by
Athena Roumboutsos,
Hans Voordijk and
Aristeidis Pantelias

Routledge
Taylor & Francis Group

LONDON AND NEW YORK

First published 2018
by Routledge
2 Park Square, Milton Park, Abingdon, Oxon OX14 4RN

and by Routledge
605 Third Avenue, New York, NY 10017

First issued in paperback 2020

Routledge is an imprint of the Taylor & Francis Group, an informa business

British Library Cataloguing-in-Publication Data
A catalogue record for this book is available from the British Library

Library of Congress Cataloging-in-Publication Data
Names: Roumboutsos, Athena, editor. | Voordijk, Hans, editor. | Pantelias, Aristeidis, editor.
Title: Funding and financing transport infrastructure : business models to enhance and enable
financing of infrastructure in transport / edited by Athena Roumboutsos, Hans Voordijk, and
Aristeidis Pantelias.
Description: Abingdon, Oxon ; New York, NY : Routledge, 2018. |
Series: Spon research, ISSN 1940-7653 | Includes bibliographical references and index.
Identifiers: LCCN 2017015299 | ISBN 9781138293892 (hardback : alk. paper) |
ISBN 9781315231822 (ebook : alk. paper)
Subjects: LCSH: Transportation—European Union countries—Finance. |
Transportation and state—European Union countries. | Infrastructure (Economics)—
European Union countries—Finance.
Classification: LCC HE242 .F86 2018 | DDC 388.068/1—dc23
LC record available at https://lccn.loc.gov/2017015299

ISBN 13: 978-0-367-73579-1 (pbk)
ISBN 13: 978-1-138-29389-2 (hbk)

Typeset in Goudy
by Florence Production Ltd, Stoodleigh, Devon, UK

Contents

Figures

Tables

Contributors

Joao Bernadino, TIS, Portugal

Javier Campos, University of Las Palmas de Grand Canaria, Spain

Ibsen Chivata Cardenas, University of Twente, The Netherlands

Jelena Cirilovic, University of Belgrade, Serbia

María Manuela González-Serrano, University of Las Palmas De Grand Canaria, Spain

Federico Inchausti-Sintes, University of Las Palmas de Grand Canaria, Spain

Nenad Ivanisevic, University of Belgrade, Serbia

Iosif Karousos, University of the Aegean, Greece

Agnieszka Łukasiewicz, Road and Bridge Research Institute, Poland

Miljan Mikić, University of Belgrade, Serbia

Kay Mitusch, Karlsruhe Institute of Technology, Germany

Goran Mladenović, University of Belgrade, Serbia

Panayota Moraiti, University of the Aegean, Greece

Eleni Moschouli, University of Antwerp, Belgium

Pierre Nouaille, Cerema, France

Aristeidis Pantelias, University College London, UK

Athena Roumboutsos, University of the Aegean, Greece

Murwantara Soecipto, University of Antwerp, Belgium

Alenka Temeljotov Salaj, University College of Applied Sciences, Norway

Lourdes Trujillo, University of Las Palmas de Grand Canaria, Spain

Nevena Vajdic, University of Belgrade, Serbia

Thierry Vanelslander, University of Antwerp, Belgium

Koen Verhoest, University of Antwerp, Belgium

Hans Voordijk, University of Twente, The Netherlands

Tom Willems, University of Antwerp, Belgium

Petr Witz, Charles University, Czech Republic

Foreword

The Cost–Benefit Analysis (CBA) has been considered the principal tool by which to assess the viability of an infrastructure project, including transport. Value-for-Money assessments have been considered key in identifying the viability of private co-financing. Are they enough to secure project performance?

In a changing world, with limited public funding and financiers seeking opportunities, what happens next in the long life cycle of a transport infrastructure project is probably most important. The multiple effects and impacts of this changing world, when conceptualised, may provide useful answers to questions of future outcomes and strategies to secure desired performance. A Transport Infrastructure Resilience Indicator representing a project rating could then be assessed with an eye to the future.

This was the objective of the BENEFIT Horizon 2020 EC funded project and the book seeks to increase understanding of the impacts of the project setup and its environment on project performance. It represents a strategic tool for decision makers and practitioners. It is not limited to specific stakeholder views. On the contrary, it allows each stakeholder to formulate strategies based on a holistic set of potential scenarios.

By including case studies, it aims to provide both experiences and guidelines as to the design and implementation of resilient projects delivered both through traditional and Public–Private Partnership (PPP) models.

Acknowledgements

The authors and editors of this book wish to thank Panayota Moraiti and Iosif Karousos for their contribution during the preparation of the book and their assistance in reviewing and editing.

The contents of this book are partly based on research carried out within the framework of the BENEFIT (Business models for ENhancing Funding and Enabling Financing for Infrastructure in Transport) project. The BENEFIT project has received funding from the *European Union's Horizon 2020 research and innovation programme* under grant agreement no. 635973.

Abbreviations

AADT	Average Annual Daily Traffic
ADSCR	Annual Debt Service Cover Ratios
AWV	Antwerp Agency for Roads and Traffic
BAFO	Best and Final Offer
BENEFIT	Business models for ENhancing Funding and Enabling Financing for Infrastructure in Transport
BM	Business Model
BN	Bayesian Networks
CAPM	Capital Asset Pricing Model
CBA	Cost Benefit Analysis
CII	Construction Industry Institute
CSI	Cost Saving Indicator
DBFM	Design, Build, Finance, Maintain
DBFO	Design, Build, Finance, Operate
DBOM	Design, Build, Operate, Maintain
D-TIRESI	Dynamic Transport Infrastructure Resilience Indicator
EC	European Commission
EIB	European Investment Bank
EJTIR	European Journal of Transport and Infrastructure Research
ETCR	Energy, Transport and Communications Regulation indicator
FEI	Financial-Economic Indicator
FEP	Front-End-Planning
FSI	Financing Scheme Indicator
fsQCA	Fuzzy-Set Qualitative Comparative Analysis
GCI	Global Competitiveness Index
GDP	Gross Domestic Product
GFC	Global Financial Crisis
GI	Governance Indicator
HGV	Heavy Goods Vehicle
InI	Institutional Indicator
IPC	Interim Payment Certificates
IRA	Reliability/Availability Indicator
IRR	Internal Rate of Return

ITF	International Transport Forum
LOC	Level of Control
LRT	Light Rail Transit
NMCP	National Motorway Construction Programme
NOPAT	Net Operating Profit After Tax
NTP	National Transport Program
OECD	Organisation for Economic Co-operation and Development
OJEU	Official Journal of the European Union
O-TIRESI	Overall Transport Infrastructure Resilience Indicator
PATHE	Patras-Athens-Thessaloniki-Evzoni
PDRI	Project Definition Rating Index
PFI	Private Finance Initiative
PP	Project Performance
PPIAF	Public Private Infrastructure Advisory Facility
PPP	Public Private Partnership
QCA	Qualitative Comparative Analysis
RAI	Remuneration Attractiveness Indicator
RBV	Resource-Based View
REIT	Real Estate Investment Trust
RoIC	Return on Invested Capital
RO-RO	Roll-on/Roll-off
RRI	Revenue Robustness Indicator
RSI	Revenue Support Indicator
S&P	Standard and Poor's
SPV	Special Purpose Vehicle
S-TIRESI	Static Transport Infrastructure Resilience Indicator
TEN-T	Trans-European Networks-Transport
TIRESI	Transport Infrastructure Resilience Indicator
TIRESIAS	Transport Infrastructure Resilience Assessment tool
UITP	International Association of Public Transport
VfM	Value for Money
WACC	Weighted Average Cost of Capital
WEF	World Economic Forum
WGI	World (Bank) Governance Indicators

1 Introduction

Athena Roumboutsos, Hans Voordijk
and Aristeidis Pantelias

1.1 The need for a transport infrastructure delivery performance rating

Local and global infrastructure needs, the widening 'financing gap', and the increasingly important role of the private sector in the delivery of infrastructure have been at the forefront of public and private debate over the past three decades and drawn the close attention of both academia and the industry. Transport infrastructure has always been at the core of these discussions given its economic value-adding nature as a sector and also its strong traditional reliance on public procurement and delivery. The recent global financial crisis (GFC) was only the latest reminder that governments have found it increasingly difficult to maintain a high standard of transport infrastructure in a framework of tight public spending, which has continuously declined since the 1970s (Inderst, 2013). The resulting transport infrastructure investment gap has been estimated to be in the trillion USD range (World Economic Forum, 2012; Mckinsey Global Institute, 2013). Private financing of infrastructure became a popular solution from the 1990s and onwards, as the underlying delivery models did not immediately affect public spending, which was a key driver for many governments (European Investment Bank, 2005; Irwin, 2012).

Globally, a mismatch appears to exist between demand for infrastructure investment and supply of infrastructure finance (Ehlers, 2014). Considerable effort has been made to understand the barriers that create this mismatch and identify the tools/instruments to remove them. In this book, an important distinction will be made between financing and funding, as the two are not quite the same: the former corresponds to raising capital at the beginning of a project to pay for its development costs, such as construction costs; the latter corresponds to the long-term financial streams that will support the repayment of the financing of the project during its operating life and is closely related to its overall economic and financial viability. This distinction is crucial in understanding the three major barriers that have been identified in the literature to create the aforementioned mismatch:

- lack of experience and knowledge on infrastructure both from the technical perspective of project delivery as well as with respect to its characteristics as a potentially new asset class (i.e., its risk-return profile);
- regulatory restrictions in the provision of long-term finance; and
- lack of sufficiently robust and reliable funding streams for infrastructure development.

All three barriers are interrelated: the GFC has affected funding streams and has stressed returns on existing investments putting long-term financial arrangements at risk; providers of long-term finance, seeing their returns decrease and their risk increase, have reduced their tenors and overall exposure to the sector to minimise losses; losses have alarmed regulators who have, in turn, instigated stricter regulation to protect against systemic risk; regulatory restrictions imposed have further reduced the supply of long-term finance, thus making future investment in infrastructure more expensive; expensive financing makes projects more difficult to approve and requires stronger underlying funding streams in order to be viable, which goes full circle to the beginning of the discussion. Unless the vicious circle is broken, the mismatch seems to be hard to abate.

Numerous authors have identified the economic and financial characteristics that can make infrastructure investments attractive: long-term assets with long economic life; low technological risk; provision of key public services creating strong, inelastic demand; natural monopoly or quasi-monopoly market contexts; high barriers to entry; asset regulation; frequent natural (or contractual) hedge against inflation; stable, predictable operating cash flows; attractive risk-adjusted returns and low correlation with traditional asset classes and overall macro-economic cycles (Gatti, 2012; Blanc-Brude, 2013; Valila, 2015).

Transport infrastructure, in particular, is related to economic development and, therefore, creates direct and indirect beneficiaries. These beneficiaries provide direct (such as tolls, fares and other forms of user fees) and indirect (such as potentially greater tax revenues) funding streams in support of the respective investment(s). A stream of funding (expressed as project revenue) is a necessary (but not always sufficient) underlying condition for financing, whether public or private. The terms of financing depend significantly on the reliability of the funding/revenue stream, which in turn, relies on the infrastructure's 'ability' to sustainably achieve the designed (planned/anticipated) direct and indirect benefits described by its business case and wider business model. Given such potential to generate revenues, private finance and alternative project delivery models supporting the involvement of the private sector in design, construction, operation and maintenance, as well as financing, have been adopted in various parts of the world for the provision of transport infrastructure (Engel *et al.*, 2010).

However, relying solely on the private sector to deliver infrastructure would lead to suboptimal levels of investments due to its market nature (Helm, 2009). This is even more pronounced in the case of transport. Transport infrastructure may generate substantial benefits to a wide range of sectors. These are usually difficult to measure and/or monetise. Even if measuring them was possible and accurate, monetising their impact and charging for it might still be difficult or even undesirable. In this context, some form of public sector contribution is, often, expected in transport projects, such as the provision of initial (grant) financing during construction and/or subsidies during operation.

Infrastructure projects are complex and unique. Transport infrastructure projects are no exception and have been characterised by enormous cost over-runs (some transport modes more than others) due to optimism bias in their ex-ante evaluation phase, among other possible reasons (Flyvbjerg *et al.*, 2004). To competently address this complexity, developers and investors need to build respective capabilities, which is only efficient and feasible if a pipeline of projects is available.

Transport projects are usually revenue generating. As such, revenue-related risks are significant for their viability and reflect the uncertainty in predicted traffic volumes and the willingness of users to pay for the relevant service(s). Transport pricing, apart from determining the characteristics of the corresponding revenue stream, is also highly correlated to travel behaviour, social acceptance and the feeling of fairness. Pricing becomes crucial in cases of private involvement in transport infrastructure delivery as motives may differ: while the government would aim to spread the resulting benefit at a low price by allowing service access to many users, the private sector would tend to set higher prices attracting fewer users and thus reducing its operation and maintenance costs. Additionally, in cases where the private sector acts as 'owner' of the asset, it usually requires the establishment of a 'temporary monopoly'. This condition, which can result from the natural location of the asset with respect to the underlying network or can be contractually induced, may be affected by the planning of other new or upgraded transport assets/services and general spatial planning by the public sector. In some cases, these plans may work in support; in others in competition, reflecting ultimately on revenue risk.

Additionally, many transport infrastructure projects only reach forecast traffic levels and, therefore, generate healthy cash flows, many years after their commissioning, rendering demand/revenue risks during their initial phase of operation (ramp-up) notoriously high. Asset specificity and the uniqueness of provided services also make investments in transport infrastructure projects sunk in nature and subject to hold-up risks (Evenhuis and Vickerman, 2010).

Risk exposure is very different during the different phases of transport infrastructure development. Ehlers (2014) argues that each phase requires a different mix of financial instruments to cover different risk and return profiles and so attracts different types of investors. At the same time, the issue of (lack of) expertise in the sector makes some investors dependent on others. For example, banks have built great capabilities in understanding technical due diligence and monitoring progress during construction, which are recognised and valued by other potential debt investors. Banks, hence, often serve as an implicit 'insurance' to other groups of investors with fewer monitoring capabilities, thus enabling them to participate in the financing package. However, despite these capabilities and their strategy of syndication and securitisation of loans, many banks did not survive the collapse of the inter-bank lending market during the GFC (Hellowell *et al.*, 2015).

In response to the effect of the GFC, stricter international regulations were put in place introducing more stringent capital requirements for banks (Basel III), increasingly limiting their appetite and potential to provide long-term financing instruments. This regulatory intervention led project developers and sponsors to investigate the potential of other financing instruments, such as project bonds, and other sources of long-term capital, such as institutional investors[1] (Della Croce, 2011). It should be noted, however, that institutional investors' appetite for long-term infrastructure financing has also been affected by similarly adverse regulatory interventions (Solvency II). Furthermore, capital market solutions bear a significant limitation. As bond ownership is fragmented and dispersed among multiple holders whose infrastructure-specific expertise is limited or inadequate, due diligence can no longer be kept 'in-house' as in the case of banks. While certain institutional investors have the necessary skills to evaluate and monitor such investments on their own, most capital markets infrastructure investors need to rely on third-party due diligence and relevant market signals. These are provided in the form of ratings issued by specialist rating agencies. Notably, 'creditworthiness ratings' or 'credit assessments' concern the assessment of a project 'owner's likelihood of default' (including delayed payment of debt).

According to the OECD (2014), project ratings are of particular interest to most institutional investors as in order to comply with their fiduciary duties they can only invest in investment-grade securities (rating categories BBB-/BAA and above). However, data from the period 2006–2010 show that a large part of project bonds were concentrated around BBB rating (OECD, 2014). As such ratings are too close to the threshold of noninvestment grade debt (junk), and various credit enhancement solutions have been introduced, aiming to further raise the creditworthiness of such bonds to safer categories (Hellowell *et al.*, 2015).

Pantelias and Roumboutsos (2015) profess that there are two important difficulties in achieving reliable credit assessments for infrastructure projects, especially in the case of Public–Private Partnerships (PPPs): (i) the fact that the special purpose vehicles (SPV) delivering these projects are independent entities with no financial history; and (ii) that infrastructure assets have particularities that are not equivalent to typical financial investments. Credit assessments are based on bespoke methodologies that are many times asset or sector specific. In many methodologies concerning PPP projects, the focus of the assessment is mostly on the ability of the contracting (public) authority to bear the credit risk of the project as opposed to the ability of the project to achieve forecast performance.

However, before resorting to the assessment of the project 'owner's' ability to honour debt obligations or assessing their riskiness of default, it is equally important – if not more so – to assess the likelihood of the project itself achieving its target performance outcomes. These outcomes encompass economic, social, environmental, and institutional targets, with financial targets such as debt repayment, being a smaller subset of them. Consequently, while credit assessment methodologies are concerned with risk bearing, that is, with determining whether the (legal) entity that is liable to honour (pay) debt obligations will be able to do so during the life time of the project, what seems to be missing is an assessment tool that estimates the likelihood of a project delivering on its various performance targets. This different focus places the emphasis on risk management that comes a step earlier than the ability to bear risk, under the logic that various circumstances that can jeopardise the ability of a project to deliver its expected outcomes (and to repay its debt obligations as a sub-set of them) can be anticipated and potentially mitigated through managerial actions/decisions before the need exists for one of the project stakeholders to bear the financial consequences of a risk that has eventuated.

Providing such a tool in the form of a new indicator with an underlying rating system that describes the ability of a transport Infrastructure project to withstand, adjust and recover from changes within its structural elements with respect to its ability to deliver specific outcomes (such as cost and time to completion, expected traffic and expected revenue targets) has been the objective of the BENEFIT (Business models for ENhancing funding and Enabling Financing for Infrastructure in Transport) H2020 funded project.[2] This indicator is termed Transport Infrastructure Resilience Indicator (TIRESI).

1.2 Business Models for Enhancing funding and Enabling Financing of Infrastructure in Transport: introducing the system and its rating

The TIRESI stems from the consideration of transport infrastructure delivery and operation (including maintenance) as a system of interrelated elements. The system's overall behaviour depends on its constituent elements and for a project to attain pre-specified outcome targets, the risk characteristics of these elements need to be considered and understood.

A core element of the system is the *business model*. Notably, the wider and more encompassing the business model value propositions are the wider the range of anticipated performance outcomes, which may include the following:

- direct, indirect and induced economic (and financial) outcomes;
- environmental outcomes;
- social and societal outcomes;
- institutional outcomes.

These outcomes generate varying streams of revenues stemming from public sources, captured either as direct revenues (from users of the business model value proposition), or indirect revenues (through various forms of taxation of the indirect and induced economic outcomes). The synthesis of these revenues streams constitutes the *funding scheme*. Notably, the created value of all performance outcomes cannot be captured solely in monetary terms.

The generated funding scheme forms the basis for the financing of the project. Naturally, the risk profile of the financing package matches the risk profile of the funding scheme. Hence, the *financing scheme* is created, describing the mix of sources of capital invested in the project, whether for its initial development or/and its subsequent operation/maintenance.

The flexibility and opportunities for value creation and value capture within this system are limited or enabled by the *implementation context*, describing the combined effect of legislation, regulation, administrative and soft institutions (cultural, social, etc.), as well as on-going economic–financial conditions.

The characteristics of the transport mode (*transport mode typology*), that the asset under consideration belongs to, will guide the development of the system and define the boundaries and the potential of the business model.

Finally, *governance arrangements* define, describe and regulate the relations and interrelations between the various actors involved in the development of the project, as well as their range of activities.

These are the transport infrastructure delivery and operation system *elements*. All six (governance arrangements; business model; funding scheme; financing scheme; implementation context and transport mode typology) are interrelated and interact to produce the performance outcomes, which, in return, also influence system behaviour. Therefore, performance outcomes could also be considered a final system 'element'.

All elements and their interrelations are illustrated in Figure 1.1, which presents the BENEFIT Framework.

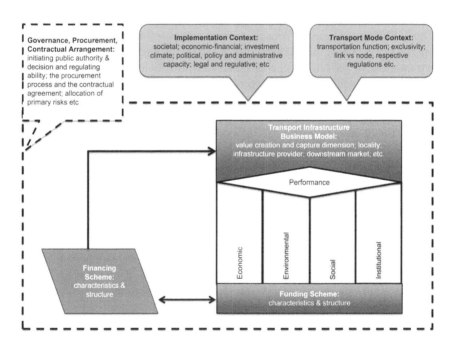

Figure 1.1 Key elements in transport infrastructure construction, operation and maintenance (BENEFIT Framework)

The BENEFIT Framework can be explained as follows:

Performance may be described by a wide range of outcomes, all of which may not be directly measurable: financial returns, economic benefits to society, environmental and institutional performance, job creation, social coherence as well as project management goals (cost and time to completion, quality) are a few to mention. Furthermore, *these outcomes are not of equal importance to all stakeholders*. The proposed TIRESI and its rating system assess the likelihood of reaching each individual performance target rather than 'overall' performance. In this context, the proposed methodology is *stakeholder neutral*, as each stakeholder can assess project performance based on their own value system by assigning respective weights to the rating of each performance target.

The *business model* and its value proposition is a key component. It is characterised by the level of integration of the project and its encompassing activities. Notably, while some characteristics will enhance the model's ability to generate revenues, other characteristics will reduce construction, operation and maintenance costs. A business model is related to strategy, value creation and value capture (Zott *et al.*, 2011) and may be employed as an 'opportunity' facilitator. In a project setting, 'value' concerns 'use-value' (Hjelmbrekke, 2017), and how specific characteristics/qualities are perceived by users in relation to their needs. Value capture is then related to the potential of use-value to generate strategic value.

The *funding scheme* describes the mix of revenue streams produced by the business model and how they are captured. These may be direct revenues; and/or investment returns (e.g. revenues generated by the use activities); and/or indirect revenues described as benefits to society. They may lead to economic growth and, ultimately, support the relevant government/public authority budget.

The *financing scheme* describes the mix of financing sources contributed for the delivery of the business model. In many ways, the funding scheme envisaged and the business model planned will guide the structuring of the financing scheme as different risk profiles, incentives and motives are put forward.

Governance describes the interaction between the formal institutions and other actors involved in the delivery of a transport infrastructure project. It is primarily based on the conditions formulating the contractual relations in the provision of the infrastructure.

Finally, all the above factors are influenced by:

- *the implementation context* that encompasses financial–economic conditions, regulations and policies describing the national conditions;
- the characteristics (typology) of the infrastructure transport mode concerned.

The BENEFIT Framework (or similar versions of it) has been studied through the collection and analysis of project cases describing the key factors influencing project performance. Notably, over the years, many assessments of investments in transport infrastructure have been carried out by and on behalf of key stakeholders (e.g. funders and financiers). Emphasis has been placed on highlighting lessons learnt, through the identification of best practices and common practices, as well as the identification of limitations to the applicability, usability and impact on performance (and on wider economic effects) of various funding and financing schemes. These assessments have provided various insights to the performance of different funding schemes but their findings have been contextually restricted. These studies were only valid within a particular institutional, micro- and macro-economic, cultural, and transport mode-related setting. Consequently, the transferability of these findings was limited, especially in light of the complexity of interactions among all these elements over the life cycle of the infrastructure.

Furthermore, over the years, many projects have been reported as successful only for this assessment to be reconsidered later in time as conditions changed and the initial project setup was not able to deliver performance as originally expected. Such changes highlight the time-related nature of these assessments considering the influence of changing conditions even in the case of the same project.

To address this shortcoming, the study of the BENEFIT Framework was conducted through the construction and elaboration of quantifiable indicators that aimed to capture the characteristics of these system elements (Roumboutsos, 2015a). This approach constitutes the BENEFIT concept and the basis of the TIRESI and its underlying rating methodology. More specifically, by capturing the combinations of indicators and their threshold values that produce a positive impact on pre-specified performance outcomes, a rating is formulated describing the likelihood of projects, described by the respective indicators, to achieve these pre-specified target outcomes.

It is important to note that the BENEFIT concept has two inherent properties. It is:

- dynamic, as elements (and their indicator values) change over time influencing the overall project performance. The dynamics of the model are expressed through the changing values of the system indicators over time. In this context, TIRESI ratings are time related;
- heuristic, in the sense that it is based on information from project cases. As such, it continuously learns from them and its predictive ability improves as the number of project cases it is based on increases.

Due to the heuristic nature of the BENEFIT concept, the proposed TIRESI rating system has been developed and calibrated based on the BENEFIT project case dataset. The dataset includes information on 86 projects delivered through traditional public financing and private (co)financing models and was populated gradually over several years by many contributing researchers (Roumboutsos et al., 2013, 2014; Roumboutsos, 2015b). The complete list of projects is included in Appendix 1.

Performance outcomes of infrastructure projects are usually defined by specialised studies, such as Cost Benefit Analysis (CBA), conducted ex-ante, that is, prior to project implementation. Their main aim is to provide justification for project delivery, that is a project 'raison d'être'. Notably, key performance outcomes include meeting cost and time to construction completion targets, and achieving traffic/demand and revenue forecasts.

The TIRESI rating system is based on the quality of the business model, the governance arrangements, the funding and financing schemes, and other factors identified as having a positive (or adverse) impact on performance. Essentially, the TIRESI rating system is based on an assessment of the 'structural' characteristics of the project.

The information that is provided by TIRESI ratings takes the form of an estimation of likelihood of achieving a selected performance target and can have multiple uses. First and foremost, TIRESI ratings are meant to support (private and/or public) investment decisions. They address the first of the three major barriers identified for the commitment of private, long-term financing to infrastructure: experience and knowledge on infrastructure characteristics. They also induce a better understanding of the third barrier, that is, the need for reliable funding streams, through the explicit connection that exists between the Funding and the Financing schemes of a project. For planners, TIRESI ratings may be of use as a tool to identify factors that would improve (or endanger) the likelihood of achieving target project performance and thus guide their inclusion (or exclusion) in (from) the project plan.

During project implementation, TIRESI ratings can also be utilised as a 'project health' monitoring tool that may allow developers/sponsors to take corrective actions, mitigate adverse effects or even assess if action is needed to be taken, improving on management efficiency.

TIRESI ratings also contribute across the board to scenario testing, especially in cases of project re-financing as well as when new financing instruments are being considered for implementation.

Finally, TIRESI ratings and their underlying methodology are equally applicable to both private (co)financed and public financed projects.

The TIRESI rating system is a novel concept that is supported by a user-friendly web-based application that computes the system indicators as well as the various performance ratings based on information provided by the user. For more information on the application please visit: www.benefit4transport.eu/benefit/.

1.3 Introducing the chapters of this book

Chapter 1 presents the basic premises behind the creation of this book. It sets the background and introduces the BENEFIT Framework, Concept, and Dataset that form the basis of all analyses undertaken and presented within the following chapters. This chapter also explains the need for the development and calibration of the new TIRESI and its rating system as a decision-support tool for various project phases and all potential project stakeholders.

Chapter 2 presents an overview of findings from a qualitative analysis conducted on the projects included in the BENEFIT project case dataset. Although the dataset contains many infrastructure projects (and their respective business models) pertaining to all transport modes, it only constitutes a small sample compared to the vast number of transport projects undertaken around the world. Consequently, a qualitative analysis aiming at the in-depth consideration of the sample and its degree of representativeness was of cardinal importance. Therefore, the chapter serves a dual objective: it presents the findings of the qualitative analysis and it compares its findings to other reported research. In this context, Chapter 2 contributes to understanding the performance of transport infrastructure projects. Furthermore, Chapter 2 demonstrates how project performance across modes may be grouped under and related to the elements of the BENEFIT Framework, thus, justifying its potential as an explanatory framework. Interrelations between factors belonging to different elements of the BENEFIT Framework are identified, but it becomes obvious that no further elaboration is possible through a qualitative approach.

Chapter 3 resolves the limitations of qualitative analysis vis-à-vis understanding and quantifying the sometimes obvious, but many times subtle, interrelationships between the elements of the BENEFIT Framework and their corresponding impact on project performance. This effort is undertaken by representing all elements through quantifiable indicators. For every element, the development, validation and operationalisation of its corresponding indicator(s) are presented. The power of representing projects through indicators is demonstrated through the elaboration of a multi-method numerical analysis of the BENEFIT project case dataset. Findings show that combinations of indicators are required to achieve project outcomes and also that their potential variations lead to an enhanced understanding of the potential approaches that may be endorsed for the attainment of project outcome targets. This enhanced understanding of indicator combinations and their impact on project performance sets the ground for the development of the BENEFIT Transport Infrastructure Resilience Indicator and rating system.

Chapter 4 presents the development and quantification of the TIRESI and its rating system. This novel indicator aims to measure how resilient a project is, on its course to achieve its intended outcomes, in the face of its internal structure as well as its external conditions. The methodological underpinnings for the TIRESI are based on a combination of scientific research and industry practice. It builds upon academic research in project and system resilience, but also leverages elements from credit assessment methodologies that have been developed and applied by specialised firms in the financial services industry. The quantification methodology of the TIRESI and the application of its rating system are developed by calibrating the findings of indicator combinations as produced in Chapter 3 against the considerations per transport mode reviewed in Chapter 2, providing a rating system per mode. This methodology provides the necessary tools for a full demonstration of its applicability that is elaborated upon in Chapter 5.

Chapter 5 presents the analysis of nine (9) project cases based on the BENEFIT Framework and the TIRESI. The nine projects belong to the following transport modes: road, urban transit, fixed link (tunnel) and port. Their analysis demonstrates the usability and applicability of the BENEFIT Framework and the indicators representing its elements in analysing and understanding project performance. The use of the TIRESI also enables the investigation of potential actions that can be taken to improve the likelihood of these projects to attain some of their outcome targets and the testing of alternative development scenarios.

Chapter 6 presents the implications of the development of the TIRESI and its rating system to decision makers. Its usefulness is considered in the context of already existing credit rating assessments, and other systems rating infrastructure delivery, which it aims to complement, as well as the needs of various project stakeholders whose project-related decisions it aims to support. This chapter suggests that TIRESI ratings can provide a valuable measure and important market signal for the potential of success of transport infrastructure projects when it comes to the attainment of specific outcome targets.

Finally, *Chapter 7* presents the conclusions from the research effort undertaken and puts forward topics for further research.

1.4 Notes

1 The term institutional investors encompasses banks, insurance companies, pension funds, hedge funds, Real Estate Investment Trusts (REITs), investment advisors, endowments and mutual funds. Considering investment in infrastructure, the terms usually refers to pension funds, insurance companies as well as sovereign wealth funds.
2 See www.benefit4transport.eu; EU Grant Agreement No. 635973.

1.5 References

Blanc-Brude, F. (2013). *Towards Efficient Benchmarks for Infrastructure Equity Investments*. EDHEC-Risk Institute Publications.

Della Croce, R. (2011). *Pension Funds Investment in Infrastructure: Policy Actions*. Paris: OECD, pp. 231–287.

Ehlers, T. (2014). Understanding the Challenges for Infrastructure Finance. *Bank of International Settlements Working Papers No. 454*. www.bis.org.

Engel, E. M. R. A., Fischer, R. D., and Galetovic, A. (2010). The Economics of Infrastructure Finance: Public-Private Partnerships versus Public Provision. *European Investment Bank Papers*, 15(1), pp. 40–69, ISSN 0257-7755.

European Investment Bank. (2005). Evaluation of PPP. Projects Financed by the EIB, Luxembourg.

Evenhuis and Vickerman (2010). Transport Pricing and Public-Private Partnerships in Theory: Issues and Suggestions. *Research in Transportation Economics*, 30, pp. 6–14.

Flyvbjerg, B., Skamris Holm, M.K., and Buhl, S. L. (2004). What Causes Cost Overrun in Transport Infrastructure Projects? *Transport Reviews*, 24(1), pp. 3–18.

Gatti, S. (2012). Fueling European Union Growth: Financing and Investing in Infrastructure, *Working Paper*, Carefin Bocconi.

Gatzert, N., and Kosub, T. (2016). The Impact of European Initiatives to Promote Infrastructure Investments from the Insurance Industry's Perspective. Working Paper (version April 2016). Friedrich-Alexander University Erlangen-Nürnberg (FAU).

Hellowell, M., Vecchi V., and Caselli S. (2015). Return of the state? An appraisal of policies to enhance access to credit for infrastructure-based PPPs. *Public Money & Management*, 35(1), pp. 71–78.

Helm, D. (2009). Infrastructure, Investment and the Economic Crisis. In D. Helm, J. Wardlaw, and B. Caldecott (eds.), *Delivering a 21st Century Infrastructure for Britain*, London: Policy Exchange.

Hjelmbrekke, H., Klakegg O. J., and Lohne, J. (2017). Governing Value Creation in Construction Project: A New Model. *International Journal of Managing Projects in Business*, 10(1), pp. 60–83.

Inderst, G. (2013). Private Infrastructure Finance and Investment in Europe. *EIB Working Papers*, 2013/02.

Irwin, T. (2012). Accounting Devices and Fiscal Illusions. *IMF Staff Discussion Note*, Washington, DC.

McKinsey Global Institute. (2013). Infrastructure Productivity. How to Save $1 Trillion a Year, January 2013.

OECD. (2014). Private Financing and Government Support to Promote Long-Term Investments in Infrastructure. Paris, Sept. 2014.

Pantelias, A., and Roumboutsos, A. (2015). A Conceptual Framework for Transport Infrastructure PPP Project Credit Assessments. *Journal of Finance and Economics*, 3(6), pp. 105–111.

Roumboutsos, A. (2015a). Case Studies in transport Public Private Partnerships: Transferring Lessons Learnt. *Transportation Research Record*, pp. 15–4638.

Roumboutsos, A. (2015b). *Business Models for Enhancing Funding and Enabling Financing for Infrastructure in Transport: PPP and Public Transport Infrastructure Financing Case Studies. Horizon 2020 European Commission*. Greece: Department of Shipping, Trade and Transport, University of the Aegean, ISBN 978–618–82078–1-3.

Roumboutsos, A., Farrell, S., Liyanage, C. L., and Macário, R. (2013). COST Action TU1001 Public Private Partnerships in Transport: Trends & Theory P3T3, 2013 Discussion Papers Part II Case Studies, ISBN 978–88–97781–61–5.

Roumboutsos, A., Farrell S., and Verhoest, K. (2014). COST Action TU1001 – Public Private Partnerships in Transport: Trends & Theory – 2014 Discussion Series – Country Profiles & Case Studies, ISBN 978–88–6922–009–8.

Valila, T. (2015). So What Is Infrastructure Anyway? Keynote Lecture, The Bartlett School of Construction and Project Management Keynote Lecture Series, UCL, 24 March 2015

World Economic Forum. (2012). Strategic Infrastructure, Steps to Prioritize and Deliver Infrastructure Effectively and Efficiently. www.weforum.org (Accessed September 2012).

Zott, C., Amit, R., and Massa, L. (2011). The Business Model: Recent Developments and Future Research. *Journal of Management*, 37(4), pp. 1019–1042.

2 Key mode-specific issues in funding and financing of transport infrastructure

Goran Mladenović

2.1 Introduction

This chapter presents a review of the business models underpinning a number of public and PPP transport projects included in the BENEFIT case study set. It also undertakes an assessment of their performance with respect to a number of outcomes, namely cost and time to construction completion, actual versus forecast traffic, actual versus forecast revenues as well as the achievement of other transport goals. The objective is to identify key factors leading to project success or failure with respect to these outcomes, as well as to analyse the impact of the GFC on project performance. Findings are compared to the results of previously reported case study based research in order to achieve the verification or testing of theory by the logic of replication (Eisenhardt, 1989; Taylor *et al.*, 2009). Notably, case study based research has been the preferred research tool for the study of PPPs (Tang *et al.*, 2010), as it is well situated to address 'how' or 'why' questions about a contemporary set of events, over which the investigator has little or no 'control' (Yin, 2009).

As noted in Chapter 1, the BENEFIT project case dataset (see Appendix) includes 86 cases. Fifty-five concern projects delivered through schemes falling under the general umbrella of PPPs, while the remaining 31 projects are financed by the public sector. The projects span across all transport modes, including road, bridge and tunnel (fixed link), seaport, airport, rail and urban transit, in 18 European countries.

Transport infrastructure modes considered in this study differ in many respects, considering their physical characteristics (nodal, such as seaports and airports; linear, such as roads and rail; network, such as urban transit networks), the relative importance of construction and operation phases, the business models employed, etc. However, they are all analysed based on the BENEFIT Framework (see Chapter 1) describing the transport infrastructure delivery system. This framework endorses Stake's (2000) view of a case being considered as an 'integrated system' with an identity, purpose and working parts, and provides the relevant 'context' for analysis across all case studies.

The chapter is structured addressing each transport infrastructure mode separately. Within each mode, key factors influencing performance are presented.

As discussed in the concluding part of this chapter, some identified factors are significant across all modes, while others bear less importance. As mentioned earlier, when considering performance, four key outcomes are being considered throughout the analysis. Two of them pertain to project management goals: cost and time to completion. One is directly related to the basic justification of any transport infrastructure: traffic. The last, but certainly not the least, is related to economic performance: project revenues.

2.2 Road infrastructure

The road network forms the core element of a country's transport system providing mobility for people and goods. A reliable road network takes years to develop and requires timely and cost-effective maintenance.

Several western European countries, such as the United Kingdom, Spain, Italy, and France, have successfully employed private financing to deliver road infrastructure. In Scandinavian countries (e.g. Norway and Sweden), private participation in such projects has been limited, while substantial potential exists. In central and eastern Europe, mixed experience has been witnessed in countries such as Hungary, Croatia, and Poland (Queiroz *et al.*, 2016). However, in certain southern European countries (e.g. Greece and Portugal), many PPP road projects faced significant viability problems following the GFC.

The sample of road projects analysed within the BENEFIT project case dataset included 31 cases. Twenty-one cases were co-financed by the private sector (PPP delivery type) and 7 were financed through the public budget. The projects are located in 15 European countries and were awarded between 1987 and 2012. Most of the PPP cases in the sample were awarded prior to the GFC, while five of seven public financed cases following the crisis (awarded between 2009 and 2011). The sample included 13 greenfield projects, 8 brownfield projects and 10 projects that contained both greenfield and brownfield sections (Mladenovic *et al.*, 2016; Roumboutsos *et al.*, 2016a).

The analysis has identified several issues that are characteristic of road infrastructure:

- the importance of the project implementation context;
- cost overrun and delays in project implementation;
- traffic forecasting performance;
- limited ability to support their business model;
- vulnerability to exogenous adverse impacts.

2.2.1 *The importance of the project implementation context*

There is substantial difference in performance between projects in northern and western European countries, and projects in southern European countries. Most projects that experienced cost and time overrun and traffic overestimation are located in southern countries (Greece, Italy, Serbia and Spain). Only one project in the sample – the Combiplan Nijverdal project, located in the Netherlands – does not follow the above trend. Projects in Belgium, Croatia, Finland, Germany, Norway and Poland are performing in line with expectations (or better) regarding cost and time to construction completion as well as traffic forecast. Notably, in the United Kingdom, where there was significant impact from the GFC, the M-25 Motorway London Orbital project (under construction at the time) was influenced. Finally, the only UK project adopting real tolls as a remuneration scheme (BNRR – M6 Toll road) has actual traffic below initial forecast.

Canterelli *et al.* (2012) found that project performance regarding cost overrun varies with location, and that for transport infrastructure projects located in the Netherlands, the cost overrun is smaller compared to the rest of the world. However, their conclusion for road and tunnel projects was that the cost overrun was similar to the rest of the world, while for bridges it was substantially smaller – the mean cost overrun was 6.6 per cent for seven projects in the Netherlands compared to 45.1 per cent for 15 projects delivered in Switzerland, Germany, Denmark, France, Norway, Sweden, United Kingdom and Hungary.

Notably, variation in performance is related to the maturity of institutions and the stability of the macro-economic environment in each country, as these favour the development and implementation of mature and successful projects. Good governance during project construction seems to be critical for project success too, which is also related to the maturity of institutions.

2.2.2 Cost overrun and delays in project implementation

The analysis within the BENEFIT project case dataset (Mladenovic *et al.*, 2016, pp. 24–46; Roumboutsos *et al.*, 2016a, pp. 33–46) found that the cost overrun was mainly present in medium-sized projects, that is projects with construction budgets ranging between 400 and 1000 M EUR. Five of 11 (45 per cent) medium-sized projects experienced cost overrun. The percentage was around or below 20 per cent for both small and large projects. Other studies, such as the one conducted by Odeck (2004), found that cost overrun appears to be more predominant among smaller projects compared to larger ones. Hence, these findings may be sample specific.

More than half of the PPP projects awarded between 1999 and 2003 experienced cost overrun, as well as two public projects that were awarded in 2009 and 2010, respectively, following the GFC. The study also showed that brownfield PPP and greenfield public road projects are delivered more successfully regarding cost criteria, since there was no cost overrun for these two categories.

The main reasons for cost overrun in road projects are typically related to scope changes, expropriation and archaeology problems, and, in some cases, to other technical issues (e.g. Moreas Motorway and Combiplan-Nijverdal). The most typical reasons for delays in road projects are expropriation problems, design changes, technical and archaeology issues, but, also, problems related to the status and competence of the contractor (e.g. bankruptcy of the contractor in the Koper-Izola Motorway). Notably, such problems do not always have a negative impact. For example, on the Horgos-Novi Sad Motorway, the contractor went into bankruptcy but this did not effect on project outcomes. The economic crisis was, also, found to negatively impact time to construction completion.

2.2.3 Traffic forecasting performance

Based on the analysis of 104 privately financed road, bridge and tunnel projects, Bain (2009) found that traffic performance ranged between 0.14 and 1.51 (i.e. the actual traffic ranged between 86 per cent below forecast to 51 per cent above forecast). The mean value was 0.77, meaning that, on average, the respective forecasts were optimistic by 23 per cent. The main reported drivers of forecast error included recession or economic downturn, but also lower than expected time savings, over-estimated drivers' willingness to pay tolls, improvements to competitive (toll-free) routes, low off-peak or weekend traffic (periods not often modelled in detail) and truckers' resistance to paying tolls. A similar study, performed on 183 road projects out of which more than 90 per cent were toll-free, found the mean value of traffic performance to be 0.96 (Bain and Plantagie, 2004), indicating the impact road charging may have on achieving traffic forecasts.

The analysis of the BENEFIT project case dataset (Mladenovic *et al.*, 2016, pp. 32–45; Roumboutsos *et al.*, 2016a, pp. 40–43) found that traffic forecasts are typically more optimistic for PPP projects. This is consistent with many other studies that have shown that traffic forecasts for tolled roads tend to be more optimistic and, in general, poorer than those for toll-free roads. The study also found that the GFC had significant impact on traffic levels, particularly on projects in southern European countries, where the impact of the crisis was stronger and more prolonged (see Figure 3.2, Chapter 3) with traffic levels decreasing in most projects due to this effect. In other cases (e.g. A-22 and A-23 in Portugal), the introduction of real tolls following renegotiations had a negative impact on traffic levels achieved. However, some sound projects were able to return to traffic levels before the crisis (see M-45 in Spain and E-4 in Finland). Moreover, when demand risk allocation was considered, it was identified that optimal risk allocation also leads to improved performance with respect to traffic (Roumboutsos *et al.*, 2016a, pp. 246–248). This is also supported by Bain's (2009) findings, as in many cases, demand risk is inappropriately passed to the private sector (Roumboutsos and Pantelias, 2015).

2.2.4 Limited ability to support their business model

Key characteristics that might improve project resilience (exclusivity, multiple revenues streams, etc.) have limited applicability in the case of road infrastructure (Roumboutsos *et al.*, 2016b). In addition, in Europe, the development of road infrastructure in most regions is such that new projects need special consideration as to their value propositions. More specifically:

- New projects addressing missing and cross-border links, while bearing similarities with respect to key project characteristics, need to address special issues of governance, and institutional and public authority competence, among others.
- New regional road financing should carefully consider potential value-adding propositions including the possibility of their inclusion in wider development projects.
- Finally, financing schemes geared towards maintenance and improvement projects (brownfield projects) need to be considered. Including innovation (new technologies, materials, etc.) is highly important. The potential of including innovation providers as project sponsors should be investigated.

Notably, in the BENEFIT project case dataset, none of the road projects achieved the maximum potential of their business models (Roumboutsos *et al.*, 2016b).

2.2.5 *Vulnerability to exogenous adverse impacts*

Road infrastructure is the most vulnerable to exogenous adverse impacts. The qualitative analysis of the road cases showed that most projects in countries in which the crisis impact was low to moderate[1] are performing well. The consequences of the GFC may be most clearly seen in the cluster of countries with a high impact from the crisis. The latter includes mostly southern European countries and the United Kingdom. However, most of the UK projects are performing in line with expectations and do not appear to be severely impacted by the crisis, stressing the importance of strong institutional context and proper project selection and, also, the fact that they are toll-free (with the exception of the BNRR road project). In addition, most UK road projects had reached a mature operation level before the GFC. In countries experiencing a greater impact from the GFC, most projects faced difficulties that may also indicate that poor project selection and a weak institutional context might have rendered these countries more vulnerable to the crisis. However, there are some projects that are still performing reasonably well (e.g. Athens ring road in Greece, A5 Maribor-Pince motorway in Slovenia, and M-45 in Spain) despite the lasting and continuing effects of the GFC on macro-economic conditions (see Figure 3.2, Chapter 3).

The critical success factors that led to the improved performance of these projects include:

- long-term planning;
- being top priority projects;
- having realistic traffic projections;
- being medium size projects;
- having strong regulatory body and governmental support;
- having a responsible and well-experienced concessioner;
- introducing innovations (Norway, Poland, United Kingdom).

The consequences of the GFC on the poorly performing projects were reflected through:

- contract renegotiations (Portugal, Greece, United Kingdom);
- reduction in project scope (some being rather significant as in Greece);
- increase in government co-financing (Greece);
- introduction/increase of user-paid tolls (Portugal, Greece);
- payment of claims (Greece);
- cost underestimations (Serbia);
- time overruns (Serbia, Greece);
- substantial drop in reported traffic and revenues (Spain, Portugal, Greece);
- inappropriate risk allocation between the public and private partner (Greece);
- cash flow slowdown because of public budget restrictions (Spain, Greece, Portugal).

With respect to new projects awarded following the GFC, a key observation concerns the remuneration scheme, typically demand and user based for road projects. The changes in the adopted remuneration scheme following the crisis were investigated for projects in Europe and Latin America. The analysis was based on external databases provided by the OECD and the World Bank's PPIAF. In Europe, there is a clear shift toward availability-based remuneration schemes that may be the result of PPP sponsors' and lenders' risk averseness, with the upside effect of generally being a more appropriate demand risk allocation for road projects (Roumboutsos and Pantelias, 2015). This may also be a result of the European sovereign debt crisis that extended until 2012, especially in southern European countries. In Latin America, there was an increased number of projects with government support between 2007 and 2010. However, the number of these projects quickly declined and most Latin American projects awarded since 2011 are demand based. Nevertheless, the low network density and, therefore, the relative exclusivity of new road projects in Latin America provide the grounds for viable demand-based remuneration schemes.

In conclusion, stemming from the relatively limited sample of projects considered, it seems that following the GFC, two major changes have taken place:

1   a shift from demand-based remuneration schemes to availability-based ones;
2   a significant change in the countries driving the PPP market in Europe and development in the transport sector (Mladenovic et al., 2016, pp. 208–212).

2.3 Bridge and tunnel infrastructure

Agnieszka Łukasiewicz

Bridge and tunnel projects are, by nature, technically demanding, but also carry the potential of demonstrating performance enhancing characteristics (exclusivity, network conductivity, etc.). Hence, apart from the general recommendations with respect to planning, competences and other important factors, their positioning in the network is of great importance as it would allow (or not) projects to take advantage of such performance-enhancing characteristics. The latter make these projects favourable candidates for private financing despite their potentially adverse remuneration scheme (usually tolls).

The analysis of bridge and tunnel projects within the BENEFIT project case dataset (see Appendix; Mladenovic *et al.*, 2016, pp. 67–91; Roumboutsos *et al.*, 2016a, pp. 58–61) identified that these projects are as follows:

- natural monopolies (most cases);
- complex projects that require long planning periods that may work both in favour and against project performance;
- cost-intensive, with unit cost multiple times higher than the cost of adjacent road or railway;
- involving international actors during construction and/or operation due to their demanding technical nature and, finally;
- creating considerable economic impacts and generating political sensitivity requiring political support.

2.3.1 Natural monopolies

Bridges and tunnels are physical network integration links. They integrate previously separate parts of the transport network, rendering them inherently exclusive and monopolistic in nature as they facilitate the crossing of natural barriers.

Some of the projects analysed (Berlin Tiergarten Tunnel, Sodra Lanken or Blanka Tunnel) provide the infrastructure to enhance mobility in large metropolitan areas, improving vehicle traffic (in the case of Tiergarten also accommodating trains and metros) between city districts.

In many cases, concession agreements strengthen this natural monopoly by limiting the influence of competing projects. Examples are the limitations in ferry licences for the Rion-Antirion crossing (Greece) in favour of the respective bridge concession project, and the Lusoponte concession (Portugal), which apart from the new Vasco da Gama Bridge also included the operation of the existing 4th Avril bridge over the Tagos river.

Their natural monopoly status along with usual travel time savings (e.g. Rion-Antirion Bridge 40', Lusoponte – Vasco da Gama Bridge 20', Millau Viaduct 20', Øresund link over 1 hour, Berlin Tiergarten Tunnel 15–30' for rail trip) make bridges and tunnels favourable PPP candidate projects. Notably, their unique position in the transport network allows bridges and tunnels to sustain both traffic and revenues even under adverse economic conditions as they address 'captive' demand.

Finally, bridge and tunnel PPP projects have usually showed conservative traffic estimates as observed traffic was greater than expected regardless of the remuneration model used, which is commonly demand-based (tolls). This was also identified in the BENEFIT project case dataset: apart from the Herren Tunnel, all cases demonstrated an exclusive character, allowing users to save travel time. Herren Tunnel did not achieve its initial estimates (traffic estimated 37,000 vs. actual 22,000 vehicles per day), as it competed against a non-tolled route with a comparable travel time (5 km or 4 minutes longer).

2.3.2 Planning period

Both the size of the projects and their high associated cost result in long to very long preparation periods (ranging from 17 to even 100 years). The analysed projects were characterised by a relatively long duration of the period from conception to completion. Designing and constructing special structures requires good preparation.

Moreover, the political sensitivity of such projects has a huge repercussion on timing. Change of plans regarding sources of financing, unrealistic expectations for market competition, and other such issues often result in a decade of delays in project implementation. The analysed projects are characterised by a relatively long period between conception and completion (Millau Viaduct: 17 years, the Øresund link: 44 years, Blanka Tunnel: 23 years, Rio-Antirio bridge: 100 years), while only the Lusoponte – Vasco da Gama Bridge was completed in the relatively short period of 6 years.

However, a long planning period, while essential, may also be a source of uncertainty and additional risk. As the planning period may span over decades, new technologies are developed, legislation might change and so may stakeholder expectations.

Bridge and tunnel projects require very large budgets, often billions of Euros; and the involvement of many stakeholders. Therefore, the complexity of implementing such projects is high. Investing time in the planning phase also allows for better assessments of the proposed solution. For example, the Viaduc de Millau is cheaper than the alternative of tunnelling through the hills flanking the river, and shortens the journey by 100 km and by up to 4 hours in the holiday season, while it also reduces pollution caused by recurrent traffic jams for the local inhabitants in Millau.

2.3.3 Cost intensity

Estimating the cost of bridge and especially tunnel projects during the feasibility stage is highly complex and challenging for state/federal agencies. The use of traditional methods to estimate tunnel/bridge project costs has led to significant cost underestimation due to limited information/data to compare different alternatives. Membah and Asa (2015) identified five principle factors contributing to cost underestimation: engineering and construction complexities, geological conditions, insufficient cost estimation, market conditions and environmental requirements. Preparing initial estimates is a challenging task especially for local authorities and a main reason why these structures are usually addressed by state/federal agencies.

Flyvbjerg *et al.* (2002) argue that cost underestimation is substantially lower for roads than for rail, bridges, and tunnels. Moreover, cost underestimation appears to be larger for tunnels than for bridges. These results suggest that the complexities of technology and geology might influence cost estimates. Flyvbjerg *et al.* (2004) conducted a statistical analysis of cost overruns and concluded that for bridges and tunnels, the available data support the claim that larger projects have larger cost escalations, whereas this does not appear to be the case for road and rail projects. For fixed link projects (tunnels and bridges), actual costs are on average 34 per cent higher than estimated costs.

Nevertheless, it was found that the projects planned and financed by a private party (PPP model) are planned more realistically in terms of deadlines of construction, and more prudently in terms of construction costs. In addition, most of the PPP projects in this group required public contribution and the support of loans granted by the European Investment Bank, which may be an indication of higher scrutiny at the front end. Moreover, bridge and tunnel PPP projects are usually undertaken by large international actors (see following section), who might be able to exploit their greater international experience.

2.3.4 Involvement of international actors

As bridge and tunnel projects are complex and technically demanding, international construction companies with relevant expertise are usually involved. This is also due to market concentration in the sector (Suarez-Aleman *et al.*, 2015; Roumboutsos *et al.*, 2017) coupled with the fact that the cost of these projects is high and only project sponsors with strong financial capacity are able to address them.

Notably, infrastructure developers, contractors and operators of the BENEFIT project case dataset of bridge and tunnel projects included Vinci (Rio-Antirio Bridge; Lusoponte, Coen Tunnel), Eiffage (Viaduc de Millau), Hochtief AG (Herrentunnel Lübeck), Bilfinger Berger (Herrentunnel Lübeck), specialised project sponsors such as Macquarie Infrastructure Corporation (Luposonte Bridge) as well as large national actors such as the Portuguese Mota-Engil SGPS (Luposonte Bridge). These actors, apart from carrying the required technical expertise, also carry the risk of their international activities. Under conditions of international growth, they come with an internationally balanced portfolio of projects that protects them from non-systemic downside risks.

However, as international players, they are vulnerable to systemic economic risks and the GFC had a significant adverse impact on company share values. For example:

- The sell-off value of Vinci decreased three times during the crisis.
- Eiffage shares continued to decline until 2011 reaching a low of 12 EUR per share before beginning a slow increase in 2012 to reach the current level of 67 EUR per share (April 2016).
- Hochtief and Bilfinger Berger shares collapsed and while Bilfinger Berger has managed to recover, Hochtief has not regained neither share value nor stability.

- The share price of the Portuguese company Mota-Engil closely corresponded to the condition of the entire Portuguese economy: The value of the company's shares fell from 8 EUR to 1 EUR and has still not reached the value recorded before the crisis.
- Macquarie Infrastructure Corp stock rate fell from 40 USD to 0.79 USD, or by 98 per cent.

The above demonstrate the potential risk associated with international actors, especially as the GFC had a greater impact on them than on most national economies (Mladenovic et al., 2016, pp. 90–91).

Finally, bridge and tunnel projects, while situated as natural monopolies, still bear the influence of changes to their respective linear infrastructure networks (road or rail).

2.3.5 Political support

Investments in transport infrastructure projects are important for local development, improving the attractiveness of real estate in the area, or reducing congestion, among other transport-related benefits. However, it is difficult to identify direct, broader economic effects resulting from these projects, besides those recognised in typical economic appraisal methodologies (e.g. Cost Benefit Analysis), such as saving time, reducing pollution, accident costs, etc. In most cases, the scale of impact is too small to affect the development of entire regions. Nonetheless, stemming from infrastructure spatial positioning, it is important to note that bridge and tunnel projects connect or facilitate connections with previously 'isolated' localities/regions and, in this context, they may have a great impact on local and regional development rendering them 'politically sensitive'. Wider policy planning and development competences are, therefore, needed when addressing the delivery of bridge and tunnel projects as well as the ability to manage competing local and regional interests. For cross-border projects, the above-mentioned considerations might also be relevant.

For example, although Environmental Impact Assessments and Cost Benefit Analyses were conducted for the Sodra Laken project (Sweden), public opposition and the lack of approved designs for toll stations delayed progress. For the Berlin Tiergarten (Germany), the main objective of the project was to improve the city's transport infrastructure to cope with expected increases in traffic volumes, and to integrate the railway into the national and European network. For the city government, it also provided an opportunity to reduce traffic and improve the urban environment in the city centre. Deutsche Bahn aimed to improve journey times and reliability. The project faced public opposition and opened in 2006, several years behind schedule. The original plan also included a metro tunnel (which finally opened in 2009) and a city railway tunnel (yet to open).

Accordingly, the importance of bridge projects for regional development is confirmed by traffic statistics: 90 per cent or most journeys by Øresund trains are within the Øresund Region, where 90 per cent of passenger travel is regional.

Four of five train passengers live in Sweden and only one-fifth in Denmark. Sweden saves 178 million euro per annum in unemployment benefits because unemployed Swedes find work in Denmark. In addition, the project itself, being a convenient and relatively cheap connection between the two countries, contributed to the development of not only the labour market, but also tourism, which recorded solid annual increases (with the exception of 2008 and 2009, when a relative decrease of –3 and –5 per cent was recorded, respectively).

Through the qualitative assessment of the BENEFIT project case dataset, it can be observed that large transport projects related to special structures such as bridges and tunnels needed support on a governmental/national level for both public and private cases. Seven cases of nine analysed could be regarded as achieving outcomes. These happen to be nationally driven projects with political support on a central government level.

Two cases, the Herrentunnel Lübeck (Germany) and the Blanka Tunnel Complex (Czech Republic), could be considered as less successful. Both are regionally driven projects. The first demonstrates a usage of the tunnel in the operational phase that is much lower than forecast. The second opened with four years of delay. The original budget of the Blanka Tunnel Complex has been exceeded twice. The idea to build the Blanka Tunnel Complex has always been pursued primarily by the representatives of the City of Prague. There has been cross-party support for its realisation with the exception of the Green Party. However, the central government's involvement was considered as very limited and mainly related to regulations. Changes in regulations have partly contributed to increases in the total cost of the project. These included, additional requirements for safety measures in tunnels by the State Mining Administration (SMA) that played a particularly important role and also fined the contractor for landslides during the construction works.

2.4 Urban transit

Pierre Nouaille

Urban transit projects are characterised by a high level of complexity that is reflected in the low degree of achievement of project outcomes. They are also subject to strong fluctuations with respect to the elements of the BENEFIT Framework during both their construction and their operation phases. The sample of urban transport projects analysed from the BENEFIT project case dataset remains small (13 case studies, see Appendix) for one to claim to be statistically representative. Nevertheless, this limited sample allows an observation of the main specific features of this mode and the issues related to their financing and funding. For example, among the 13 projects in the sample, only three have met their objectives on the four main outcomes assessed, that is construction costs, commissioning deadline, achievement of revenues and ridership forecasts (Mladenovic et al., 2016, pp. 47–66; Roumboutsos et al., 2016a, pp. 47–54).

In general, the following issues or limitations were identified in urban transport projects analysed within the BENEFIT project case study set that heavily affect their development. These projects are characterised by the:

- local dimension of their project management;
- systematic low levels of profitability; and
- requirement of permanent adjustments during their operating phases.

These limitations of urban transit project delivery are further discussed in the following three sections.

2.4.1 Local transport authorities

Urban transport projects are, in most cases, decided, designed, constructed and operated under the supervision of local contracting authorities that have a high degree of independence from national governments or from other public bodies. This situation is hardly ever met, for example, in the case of railway or road projects. Academic literature (Saussier, 2015) frequently points out the risk for these small local authorities not having the same level of competence and expertise as a State authority and being, consequently, in a weak position with respect to their private sector counterparts.

To some extent, the cases studied in the BENEFIT project case dataset highlight this difficulty, as they frequently constitute a first experience for transport authorities in the construction of metro lines (e.g. Porto metro), tramway lines (e.g. Reims tramway, Caen TVR, the Athens Tramway, etc.) or free bike sharing systems (e.g. Sevici in Seville, Velo'V in Lyon). These local transport authorities may, therefore, have a limited experience that can justify their frequent recourse to private delegation. In fact, in the present sample, 9 of the 13 cases are based on PPP contracts. The lack of expertise is, indeed, a general justification given to the solicitation of private partners and to the transfer of the associated risks.

Iossa et al. (2013), however, underlined the ambiguity of this situation: the use of PPP contracts for reasons of lack of expertise makes sense only when the contract provides a sufficient level of flexibility, allowing private shareholders to innovate. Such contracts require, in return, some expertise from the contracting authorities at least during their elaboration phase (Amaral and Saussier, 2013).

Independently from the issue of competence of the contracting authority, the local dimension of urban public transport implies a higher political sensitivity. Elected representatives, close to public transport users, are in fact highly dependent on their overall success. If service disruptions, or even a lower than expected quality of service, are usually difficult to admit for contracting authorities, it becomes unbearable in the case of urban transit projects. Such is also the case for fare increases, which are truly impossible at the end of an electoral term (Engel and Galetovic, 2014). This political sensitivity leads, in the case of long-term contracts, to 'bilateral monopoly' phenomena in which the contracting authority may be held hostage by its private delegate (Renda and Schrefler, 2006). In these situations, the private party has the opportunity to renegotiate favourable contract conditions in the absence of other competitors, which significantly reduces the value for money achieved by the public contracting authority. In these situations, risk allocation to the private party, as defined in the contract, is ineffective and this phenomenon can be described as 'risk allocation creep'.

This political sensitivity is even more important because for some projects analysed the public sector's involvement has proved to be ambiguous. In certain PPP projects, the involvement of the private shareholders has been very limited because:

- The so-called 'private consortium' is, in fact, made up of public authorities. For example, in the case of Porto metro (Portugal), the concessioner (Metro do Porto SA) consists only of public partners (Municipality of Porto, Portuguese State, Portuguese National Railway Company and the public transport company of Porto). In the case of the Brabo 1 project (Belgium), the concessioner is a private consortium consisting of both private and public shareholders.
- The PPP contract is not correlated with private financing. For instance, the first phase of the Manchester Metrolink project (UK) is based on a private concession, while the financing of the project remains exclusively public (19 per cent of debt contracted through the European Union and 81 per cent of subsidies provided by the British government).

Thus, the BENEFIT project case dataset highlights that political sensitivity, reinforced by the sometimes ambiguous interface between public and private concessioners in the case of several projects, has led to renegotiations that have regularly resulted to the detriment of the contracting authority. In such cases, public transport authorities facing renegotiations in different case studies are practically left with two options:

- increase their level of financial contribution in case of construction cost overruns (e.g. Malaga metro, Spain);
- increase the level of their operating subsidies (e.g. Porto metro, Portugal) or accepting a lower level of service when operation revenues are below expectations (e.g. Reims tramway, France).

2.4.2 Low level of profitability

The opportunistic behaviour of private partners (Beuve *et al.*, 2015) or the lack of expertise do not seem to be sufficient to explain renegotiations at the expense of public authorities. They probably also result from the inherent complexity of urban transport projects (Brown and Potoski, 2005). This complexity appears from the construction phase. Urban public transport projects are, in fact, built mainly in:

- areas with high density, which lead to difficult expropriations (e.g. Athens metro, Greece);
- territories subject to other constraints that are difficult to consider during the preliminary studies (e.g. archaeological risks, etc.), as in the cases of the Athens tramway (Greece) or Warsaw metro (Poland).

An analysis led by Flyvbjerg et al. (2002), based on a sample of 258 projects classified into three categories (rail, including urban rail projects, fixed-link, and road), aimed at studying gaps between actual and estimated costs in transport infrastructure. This study concluded that rail projects incurred the highest difference between actual and estimated costs; high-speed rail projects topped the list followed by urban rail projects.

The complexity faced by urban transport projects is, however, more obvious during the operation phase. Most of the projects studied have to face severe difficulties in the implementation of their planned funding scheme. All funding schemes within the sample studied are quite similar. They are, in principle, based on two main sources of funding: fare revenues (user charges) and public subsidies. In most cases, the operator keeps the fare revenues, which constitute the 'risky' part of their remuneration. Urban *public* transport systems are systematically making *losses as* commercial revenues are, at best, sufficient to cover only the operation costs of the project. Public transport authorities need to mitigate the lack of commercial revenues with:

- operating subsidies, which aim at compensating the difference between fare revenues and operation costs in the case of projects with a separated contract for construction (design and build contracts) and operation (operate and maintenance contracts);
- capital grants and operating subsidies, which aim at compensating both operating costs and network infrastructure depreciation charges in the case of design, build, operate and maintain (DBOM) contracts.

In both situations, estimating the necessary subsidy requires the capability to systematically predict the commercial revenues through a transport demand assessment. Problems with the level of *accuracy* of travel demand forecasts, particularly over the long term, are not specific to urban public transport (Iossa et al., 2013). However, even predicting initial use and growth rates seems rather impossible in practice (Engel and Galetovic, 2014). In all project cases analysed, apart from the two bike-sharing projects, ridership forecasts appeared to be overestimated. In fact, 6 cases of 13 show actual ridership outcomes far below expectations (i.e. Reims tramway, Porto metro, Sul do Tejo metro, Athens tramway, Malaga metro and Caen TVR). This tendency to overestimate ridership is well defined in academic literature (Pickrell, 1990).

2.4.3 Project complexity and the need for adjustments during operation

Urban transport projects, even more so than road or other fixed-link infrastructure projects, are impossible to design in isolation. They constitute necessarily links in a network made up by structural lines (partly replaced by the new project lines) and complementary lines. The integration between a restructured bus network and new tramway/metro lines constitutes a fundamental factor for achieving the ridership forecasts (Pickrell, 1990; Cerema, 2014). Setting a transit service plan including the new lines and the whole network remains complex and can only be achieved through regular adjustments.

These adjustments strongly impact the funding scheme of the public transport network and may affect:

- only the new main lines that generally use dedicated lanes (variations of frequency, adaptation of timetables, etc.);
- the entire public transport network (change in routes of feeder bus lines, evolution of fares, adaptations of timetables, etc.).

All these adjustments are, depending on the case, either in the public transport authority's hands or in the concessioner's hands or in several operators' hands. The qualitative analysis showed that these adjustments were generally not planed at construction contract award. So, in cases of 'integrated' contracts (i.e. DBOM contracts), which bundle construction and operation, operation conditions are designed very theoretically. This lack of accuracy leads to:

- renegotiations soon after commissioning (e.g. Reims tramway, France);
- premature contract terminations (e.g. stages 1 and 2 of Manchester Metrolink, UK).

More generally, long-term contracts in public transport rarely seem successful in the sample studied.

Finally, urban transport projects have various objectives (Engel and Galetovic, 2014), which are frequently different from merely achieving high financial profitability or increasing the level of ridership. Connecting urban development with transport projects (UITP, 2009), increasing modal share of public transport, improving the accessibility of disadvantaged neighbourhoods, etc. are objectives of a public transport network, which can sometimes go beyond the strict financial performance logic.

Delegated operation contracts including rigid funding schemes may lead to situations where the achievement of these different objectives is incompatible with the objective of overall financial balance. In the BENEFIT case study set, few examples of this incompatibility were observed:

- Renegotiations in the case of Reims tramway (France) led the public transport authority to reduce public transport supply to maintain the financial viability of the concession contract. Through this decision, the public authority had to abandon the principles of an equitable public transport service on its entire territory.

- Phase 2 of the concession contract of Metrolink (UK) had to be interrupted prematurely. In fact, robust demand allowed the operator to raise fares significantly. Critics speculated that these adaptations attempted to 'price off' demand and avoid additional rolling stock purchases. On the contrary, the contracting authority's initial objective (Greater Manchester Passenger Transport Authority) was to maximize ridership.

This discrepancy between optimising the funding scheme planned and achieving maintenance-related objectives is particularly visible in the case of private consortiums held mainly by investment companies (as in the cases of Malaga metro or Reims tramway). In fact, these consortiums combine:

- a strong drive to make profit
- a lack of understanding of the practicalities behind public transport operation.

2.5 Port infrastructure

Lourdes Trujillo, Javier Campos and Federico Inchausti-Sintes

Developing strong, well-functioning maritime transport infrastructure is a key element of economic growth. This is not only because most global trade is carried by sea, but also because ports are increasingly becoming an essential link for logistics chains and multimodal networks. Although in many places the government still retains a large share of decision and management power on this sector, PPPs have progressively become an interesting alternative to manage port operations in order to meet the challenges of the growing demand for new and better infrastructure services with limited resources (Pallis *et al.*, 2010). In fact, according to *Private Participation in Infrastructure Database* of The World Bank (online accessible at https://ppi.worldbank.org/), the private sector has been increasingly involved in the construction and operation of common-user port facilities in emerging countries and, until the end of 2015, 438 projects with private participation (most of them in Latin America and Asia) reached financial close, with investment commitments totalling more than 75 billion EUR.

In recent years, the economic literature has paid a lot of attention to private investments in ports in industrialised countries, trying to identify the issues that could explain their relevance as compared to the traditional management model. Liu (1995), for example, compared the performance of public and private ports in Britain, where extensive experience in infrastructure privatisation existed since the 1980s (Brooks and Cullinane, 2007). Similarly, Wang *et al.* (2014) analysed the relative efficiency between privatised and publicly operated US ports, whereas Aerts *et al.* (2014) or Rebollo Fuente (2009) focused on European ones. As expected in a sector of infrastructure with close ties to the public sector, institutional and country-specific factors appeared to be key elements explaining the successes and failures of port PPPs (Estache and Trujillo, 2008; Trujillo and Serebrisky, 2003). In the BENEFIT project case study set, six port infrastructure projects were extensively investigated: the enlargement of the port of Agaete in the Canary Islands (Spain), the *Terminal Muelle Costa* and the *Europe South Container Terminal* in Spain, the *Piraeus Container Terminal* in Greece and the expansion of the ports of Leixoes and Sines in Portugal.

The analysis of the BENEFIT project case dataset (Mladenovic *et al.*, 2016, pp. 107–113; Roumboutsos *et al.*, 2016a, pp. 67–69) led to the better understanding of the specificities of port infrastructure projects, which are focused on:

- the development of international trade; and
- the role of strategic agents in the shipping industry.

2.5.1 The impact of international trade

PPP port projects are relatively few. One of the most striking results that emerged from this analysis is that, as opposed to other transport infrastructure projects (with the additional exception of airports), ports' overall performance – regardless of their institutional structure – mostly relies on international trade. When the economy is flourishing, most ports perform satisfactorily, particularly in terms of funding and financing. During economic downturns, there is a larger pressure on authorities to improve handling efficiency, reduce port user fees, and optimise facilities to accommodate cargo flows. Under such conditions, PPP port projects are more likely to fail.

However, this situation affects private participation in different ways. Effectively, it is an incentive to adapt port facilities to new demand characteristics. In this sense, Medal-Bartual *et al.* (2016) found that productivity growth in Spanish ports was higher in the period of the GFC than in the previous period and ultimately intensified their investments (Cabrea *et al.*, 2015). Kalgora and Christian (2016) maintain that, to cope with the new situation, many ports have changed their strategic behaviour, leading to processes of port concentration and specialisation. In fact, economies of scale in cargo shipment have led to the emergence of a few global players in shipping, able to control the allocation of transhipment business to strategically located, well-equipped, and efficiently managed hub ports (Notteboom *et al.*, 2009).

2.5.2 The role of strategic agents in the shipping industry

The role of these strategic agents in the shipping industry adds a second mode-specific feature to port PPPs. Operators' strategic behaviour seems to influence traffic and revenue outcomes and port traffic is ultimately dependent on the international strategies of shipping lines and hinterland connections (logistic supply chains) (Trujillo and González, 2011). Thus, the success of such a project often goes beyond factors controlled by the operator and is mostly explained by external causes. For instance, in the Piraeus port in Greece, the presence of COSCO, the international Chinese shipping line, and the dominant position of this port in the Mediterranean seemed to be key in understanding the success of the concession (Psaraftis and Pallis, 2012).

In general, apart from the relevance of the international context and the role played by international shipping companies, few other mode-specific factors have been identified in the literature on port infrastructure. The prominent feature of the port cases is their uniqueness with respect to outcomes, as well as in a system far wider than that initially anticipated by the BENEFIT Framework.

2.6 Conclusions

Goran Mladenović and Athena Roumboutsos

This chapter presented the findings from the qualitative analysis conducted on the BENEFIT project case dataset. Findings per mode were reviewed vis-à-vis respective reported research and sector experience and reinforced existing knowledge. The analysis also demonstrated that the BENEFIT project case dataset presents similar characteristics with other project case samples reported in the literature.

The key characteristics and factors influencing the likelihood of achieving project outcomes were highlighted for each mode. Moreover, key findings may be summarised under the elements of the BENEFIT Framework.

2.6.1 Implementation context

The implementation context relates to two groups of factors that correspond to the existing financial economic conditions and the level of institutional maturity, respectively. These groups of factors are fully described in Chapter 3.

Financial economic conditions

Notably, traffic is a derived demand and, in this context, economic activity is responsible for traffic levels as well as the smooth construction/implementation of contracts.

Poor financial economic conditions could be identified behind poorly planned projects. Problems have manifested themselves in the form of cost overruns and construction delays for many modes (e.g. road, bridge and tunnel, urban transit), but also through the need for the contracting authority to re-finance or take charge of a project under adverse conditions.

The GFC and its effects are characteristic of the impact poor or declining financial economic conditions may have on the delivery of an infrastructure project and its operational performance. Roads have been identified as extremely vulnerable to these adverse external project conditions. Bridge and tunnel projects reflect the influence of the adjacent road and/or rail link and while they can sustain 'captive' traffic, they are vulnerable to the systemic risks faced by their project sponsors.

Institutional maturity

Institutional maturity was identified to be crucial in reaching target outcomes across all modes of transport infrastructure delivery. Many research contributions including those in the present chapter point to the importance of this element that may have different manifestations. In road infrastructure, it is related to the findings with respect to country performance as they represent different levels of institutional maturity, in bridge and tunnel projects to improved performance when initiated at government/state level, in urban transit projects to shortcomings due to the capability limitations of local public authorities. Deregulation of the port sector allowed for various levels of flexibility, where again, the institutional maturity of the respective authority has been an important driver of success.

Weak institutions seem to be related to cost and time overruns (for roads, bridge and tunnel projects, and urban transit projects), poor project planning, and over-estimated traffic forecasts (for road, bridge and tunnel, urban transit). In addition, they are also related to contract structure and contract governance.

The maturity of institutions is also very important in addressing political issues surrounding transport infrastructure project delivery. In many cases, these issues can be quite significant. In bridge and tunnel projects they concern the stakeholder interests of the "connected" localities. In urban transit projects, the proximity of the decision maker (local authority) and the electoral body (local citizens) may define relevant decisions.

2.6.2 Business model element

The business model is related to many identified factors: the perceived 'use-value', expressed among others through the positioning of the infrastructure in the transport network; the competence of the actors involved; the value streams produced and, most importantly, the willingness to pay.

Situated as natural monopolies bridge and tunnel projects are probably the best example of 'use-value' created and captured due to a project's position in the transport network, which is also closely related to 'captive traffic'. Notably, 'captive traffic' through improved 'use-value' is what many port terminal operators aimed for to mitigate the effects of the GFC through the improvement of productivity, concentration and specialisation. Roads, as noted, have a limited ability to support such business models and rely mostly on quality of service. Similar is the case for urban transit, where the quality of the urban transit system relies on the level of available options to passengers.

Notably, 'use-value' leading to 'captive traffic' provides a degree of resilience against financial economic downturns. Urban transit is, of course an exception, as the GFC has in many cases favoured public forms of transportation and modal shift.

Transport infrastructure projects present significant levels of complexity across all modes during both construction and operation. Therefore, these projects require commensurate competences in planning, design, construction and operation. Depending on the infrastructure mode, emphasis needs to be placed at the front end (i.e. on design and construction), such as in the case of roads, bridge and tunnel projects, on operation, as in the case of ports, or equally on both phases (urban transit projects). Actors involved need to demonstrate respective competences both technically and financially. Port terminal projects demonstrate this attribute to the extreme, as operators' behaviour seems to influence traffic and revenue outcomes. Finally, competence is required by the contracting authority in its role as planner, procurer and contract monitoring counterparty. The lack of competence of the contract authority may have serious adverse impacts that can be identified across all modes.

Being able to provide a wider value proposition is important, as its scope is to introduce economies of scale and added value. However, few such cases were identified, apart from airport projects, which are not included in the present overview. Exceptions are the Combiplan-Nijverdal project featuring combined construction of road and rail, the Luposonte Bridge concerning the construction of the new Vasco da Gama Bridge and including in its operation the 4th Avril Bridge crossing, and the construction of the Reims tramway with the inclusion of the entire transit network in the operation. Notably, these examples also reinforce the creation of conditions of exclusivity by the project's 'use value'

2.6.3 Governance element

The influence of governance is not always directly evident across the various infrastructure modes. In urban transit, is it identified in risk 'allocation creep', as, following the initial contract award, terms of reference, especially with respect to demand/revenue risk allocation, are gradually not honoured and effectively changed in favour of the actor creating a 'hold up' condition. The consequences of the GFC on road projects (contract renegotiations, increase in government co-financing, inappropriate risk allocation, changes in remuneration schemes and many others) are also related to contractual governance.

2.6.4 Financing and funding scheme elements

The overview presented in this chapter showed no clear differentiation between financing schemes. No clearly improved performance with respect to meeting project targets was identified if a project was delivered through a PPP model or through public financing. Interpretations may vary, ranging from the attribution of greater importance on other system elements, to the consideration of other incentives that might be induced, to the effect 'control' over operation (or construction) an operator (or builder) may have. A typical example of the last interpretation is port infrastructure with private port operators related to shipping lines (logistic chains) being able to favourably influence performance.

In cases where improved performance might be identified with respect to PPP projects, this might not be due to the financing mix but rather to the financing sources that also add an additional level of planning and project monitoring.

With respect to funding schemes, revenues are found to be closely related to the underlying business model and traffic outcomes to the conditions of the implementation context.

Remuneration schemes, however, across all modes, influence traffic levels according to their respective degree of price elasticity. They are, also representative, of the risk averseness demonstrated by private sponsors, as captured by the prevailing trend in road remuneration schemes following the GFC switching from user- to availability-based. Finally, the remuneration scheme is also demonstrative of contracting authority policy actions, as in the example of urban transit subsidies.

2.6.5 Concluding remarks

This chapter has shown that key findings influencing the potential of transport infrastructure projects to reach pre-specified project outcomes may be summarised under the BENEFIT Framework elements. These elements help set findings within a context that enhances the ability to interpret them and deepens their understanding.

It is worth noting, that during this process of contextualisation, many interrelations and interdependencies of these elements could be identified. However, the qualitative analysis of project cases has limited inductive power. Therefore, representing the BENEFIT Framework elements through quantifiable indicators intends to overcome this limitation and extend the investigation of their impact on the attainment of project performance. This process is presented and analysed in Chapter 3.

2.7 Note

1   The impact of crisis has been assessed based on the change (drop) of the Macroeconomic and FEI from the BENEFIT Implementation context following the crisis. See Chapter 3 for a detailed discussion on these indicators.

2.8 References

Aerts, G., Grage, T., Dooms, M., and Haezendonck, E. (2014). Public-Private Partnerships for the Provision of Port Infrastructure: Stakeholder Analysis of Critical Success Factors in German, Dutch and Belgian Seaports, *IAME 2014 Conference*, July 15–18 – Norfolk, VA, USA.

Bain, R. (2009). Error and Optimism Bias in Toll Road Traffic Forecasts. *Transportation*, 36(5), pp. 469–482.

Bain, R., and Plantagie J.W. (2004). *Traffic Forecasting Risk: Study Update 2004*. London: Standard & Poor's.

Beuve, J., Le Lannier, A., and Le Squeren, Z. (2015). Renégociation des contrats de PPP: risques et opportunités. In Saussier, S. (2015). *Economie des partenariats public-privé*. Louvain-la-Neuve: De Boeck Supérieur. pp. 165–192.

Brooks, M., and K. Cullinane (2007). Devolution, Port Governance and Port Performance, In: M. Brooks, K. Cullinane (Eds.), *Research on Transportation Business and Management*, 17, pp. 702.

Brown, T-L., and Potoski, M. (2005). Transactions Costs and Contracting: the Practitioner Perspective. *Public Performance & Management Review*, 28(3), pp. 326–351.

Cabrera, M. Suárez-Alemán, A. and Trujillo, L. (2015). Public Private Partnerships in Spanish Ports: Current status and future prospects. *Utilities Policy*, 32(C), pp. 1–11

Cantarelli, C. C., Flyvbjerg, B., and Buhl, S. L. (2012). Geographical Variation in Project Cost Performance: The Netherlands versus Worldwide. *Journal of Transport Geography*, 24, pp. 324–331.

CEREMA. (2014). *Projet de transport collectif en site propre (TCSP), Recommandations pour la mise en œuvre*. Bron: Editions du Cerema.

Eisenhardt, K. M. (1989). Building Theories from Case Study Research. *Academy of Management Review*, 14, pp. 532–550.

Engel, E., and Galetovic, A. (2014). Urban Transport: Can Public-Private Partnerships Work? *Policy Research Working Paper 6873*, The World Bank.

Estache, A., Trujillo, L. (2008). Privatization in Latin America: The good, the ugly and the unfair. Privatization Successes and Failures. pp. 136–169. Columbia University Press 2008.

Flyvbjerg, B., Holm, M-S., and Buhl, S. (2002). Underestimating Cost in Public Works Projects, Error or Lie? *APA Journal*, 68(3), pp. 279–291.

Flyvbjerg, B., Holm, M-S., and Buhl, S. (2004). What Causes Cost Overrun in Transport Infrastructure Projects? *Transport Reviews*, 24(1), pp. 3–18.

Iossa, E., Spagnolo, G., and Vellez, M. (2013). *The Risks and Tricks in Public-Private-Partnerships*. IEFE. www.iefe.unibocconi.it (Accessed 25 October 2016).

Kalgora, B., and Christian, T.M.2016). The Financial and Economic Crisis, Its Impacts on the Shipping Industry, Lessons to Learn. *Open Journal of Social Sciences*, 4, pp. 38–44.

Liu, Z. (1995). The Comparative Performance of Public and Private Enterprises. The Case of British Ports. *Journal of Transport Economics and Policy*, 29(3), pp. 263–274.

Medal-Bartual, M., Molinos-Senante, and Sala-Garrido, R. (2016). Productivity Change of the Spanish Port System: Impact of the Economic Crisis. *Maritime Policy & Management*, 43(6), pp. 683–705.

Membah, J., and Asa, E. (2015). Estimating Cost for Transportation Tunnel Projects: A Systematic Literature Review. *International Journal of Construction Management*, 15(3), pp. 196–218.

Mladenović, G., Roumboutsos, A., Campos, J., Cardenas, I., Cirilovic, J., Costa, J., González, M.M., Gouin, T., Hussain, O., Kapros, S., Karousos, I., Kleizen, B., Konstantinopoulos, E., Lukasiewicz, A., Macário, R., Manrique, C., Mikic, M., Moraiti, P., Moschouli, E., Nikolic, A., Nouaille, P.F., Pedro, M., Sintes, F.I., Soecipto, M., Trujillo Castellano, L., Vajdic, N., Vanelslander, T., Verhoest, C., and Voordijk, H. (2016). Deliverable D4.4-Effects of the Crisis & Recommendations, BENEFIT (Business Models for Enhancing Funding and Enabling Financing for Infrastructure in Transport) Horizon 2020 Project, Grant Agreement No. 635973. www.benefit4transport.eu/index.php/reports.

Notteboom, T., Ducruet, C., and De Langen, P.W. (2009). *Ports in Proximity: Competition and Coordination among Adjacent Seaports*. New York: Routledge.

Odeck, J. (2004). Cost Overruns in Road Construction – What Are Their Sizes and Determinants? *Transport Policy*, 11(1), pp. 43–53.

Pallis, A., Vitsounis, T.K., and De Langen, P.W. (2010). Port Economics, Policy and Management: Review of an Emerging Research Field. *Transport Reviews*, 30(1), pp. 115–161.

Pickrell, H. (1990). *Urban Rail Transit Projects: Forecast versus Actual Ridership and Cost.* U.S. Department of Transportation. 162 pp. http://people.plan.aau.dk/~mortenn/misc/alon/Pickrell%20%281990%29%20Urban%20Rail%20Transit%20Projects%20Forecast%20Versus%20Actual%20Ridership%20and%20Cost.pdf (Accessed 9 November 2016).

Psaraftis, H.N., and A. Pallis (2012). Concession of the Piraeus Container Terminal: Turbulent Times and the Quest for Competitiveness. *Maritime Policy & Management*, 39(1), pp. 37–43.

Quieroz C., Uribe, A., and Blumenfeld, D. (2016). Mechanisms for Financing Roads: A Review of International Practice, Technical Note No IDB-TN-1102, Inter-American Development Bank, October 2016.

Rebollo Fuente, A. (2009). La experiencia española en concessiones y APPs: Puertos y Aeropuertos, Programa para el Impulso de Asociaciones Publico-Privadas en Estados Mexicanos (PIAPPEM) and Multilateral Investment Fund (MIF) of the Inter-American Development Bank (IADB) (Spanish).

Renda, A., and Schrefler, L. (2006). Public-Private Partnerships, Models and Trends in the European Union. Report for the European parliament. 15 pp. www.eurosfaire.prd.fr/7pc/doc/1265964211_ppp_briefing_note_en.pdf (Accessed 25 October 2016).

Roumboutsos, A. (2015). Case studies in Transport Public Private Partnerships: Transferring Lessons Learned. *Transportation Research Record, Journal of the Transportation Research Board*, 2530, pp. 26–35.

Roumboutsos, A., Farrell, S., Liyanage, C.L., and Macário, R. (2013). COST Action TU1001. Public Private Partnerships in Transport: Trends & Theory P3T3, Discussion Papers Part II Case Studies. ISBN 978–88–97781–61–5.

Roumboutsos, A., Farrell S., and Verhoest, K. (2014). COST Action TU1001. Public Private Partnerships in Transport: Country Profiles & Case Studies. ISBN 978-88-6922-009-8.

Roumboutsos, A., and Pantelias, A. (2015). Allocating Revenue Risk in Transport Infrastructure PPP Projects: How It Matters. *Transport Reviews*, 35(2), pp. 183–203

Roumboutsos, A., Bange, C., Bernardino, J., Campos, J., Cardenas, I., Carvalho, D., Cirilovic, J., González, M.M., Gouin, T., Hussain, O., Kapros, S., Karousos, I., Kleizen, B., Konstantinopoulos, E., Leviäkangas, P., Lukasiewicz, A., Macário, R., Manrique, C., Mikic, M., Mitusch, K., Mladenović, G., Moraiti, P., Moschouli, E., Nouaille, P.F., Oliveira, M., Pantelias, A., Sintes, F.I., Soecipto, M., Trujillo Castellano, L., Vajdic, N., Vanelslander, T., Verhoest, K., Viera, J., and Voordijk, H. (2016a). Deliverable D4.2- Lessons Learned – 2nd Stage Analysis, BENEFIT (Business Models for Enhancing Funding and Enabling Financing for Infrastructure in Transport) Horizon 2020 Project, Grant Agreement No. 635973. www.benefit4transport.eu/index.php/reports.

Roumboutsos A., Gouin T., Leviäkangas P., Mladenović G., Nouaille P.-F., Voordijk J., Moraiti, P., and Cardenas, I. (2016b). Transport Infrastructure Business Models: New Sources of Funding and Financing. *Proceedings of the WCTR 2016 Conference, The 14th World Conference on Transport Research*, Shanghai, 10–14 July 2016.

Saussier, S. (2015). *Economie des partenariats public-privé*. Louvain-la-Neuve: De Boeck Supérieur.

Stake, R. E. (2002). *Case Studies*. In N. K. Denzin, Lincoln, and Yvonna S. (Ed.), *Handbook of Qualitative Research* (2nd edition, pp. 134–164). Thousand Oaks, CA: Sage Publications.

Suarez-Aleman, A., Roumboutsos, A., and Carbonara, N. (2015). The Transport PPP Market: Strategic Investors, Roumboutsos A. (2015) Public Private Partnerships in Transport: Trends and Theory. Abingdon: Routledge, ISBN: 978-1-138-90970-0.

Tang, L., Shen, Q., and Cheng, E.W. (2010). A Review of Studies on Public–Private Partnership Projects in the Construction Industry. *International Journal of Project Management*, 28(7), pp. 683–694.

Taylor, J., Dossick, C., and Garvin, M. (2009). Conducting Research with Case Studies. In *Proceedings of the 2009 Construction Research Congress*, Seattle, WA, ASCE, Reston, VA, 5–7 April, 2009.

Trujillo, L. and González, M. (2011). Maritime ports. International Handbook for the Liberalization of Infrastructures, M. Finger and R. Künneke ed., Edward Elgar.

Trujillo, L. and Serebrisky, T. (2003). Market Power in Ports – A Case Study of Post-privatization Mergers. Public Policy for the Private Sector, 260, pp. 1–4.

Wang, G.W., Knox, J.J., and Lee, P.T.W (2013). A Study of Relative Efficiency between Privatized and Publicly Operated US Ports. *Maritime Policy & Management*, 40(4), pp. 351–366.

World Economic Forum. (2014). *Strategic Infrastructure Steps to Operate and Maintain Infrastructure Efficiently and Effectively. Prepared in the Collaboration with the Boston Consulting Group.* April 2014. www3.weforum.org/docs/WEF_IU_StrategicInfrastructureSteps_Report_2014.pdf.

Yin, R. K. (2009). *Case Study Research: Design and Methods. Applied Social Research Methods* V.5 (4th ed.). Thousand Oaks, CA: SAGE Publications.

3 Transport infrastructure delivery in context

Thierry Vanelslander and
Athena Roumboutsos

3.1 Introduction

The review of the qualitative analysis of project cases presented in Chapter 2 identified that the key elements of the BENEFIT Framework are present across all transport modes. In each case, these elements are decisive with respect to the potential of reaching target outcomes to a greater or lesser degree.

Chapter 3 makes additional progress in capturing the effects of the transport delivery system elements on the attainment of target project outcomes. It elaborates on the effort to represent these elements in a quantifiable indicator form. This representation allows for further exploratory analysis to be undertaken and additional insight and understanding to be gained. Formulating these indicators starts from theoretical concepts found in previously reported research. Indicators are validated as per their applicability to and explanatory value towards the system framework by utilising the BENEFIT project case dataset. As a last step, indicators are operationalised by determining appropriate mathematical formulations (where appropriate). All indicators (with one exception) are constructed so that they take values in the range [0,1]. Higher indicator values reflect project characteristics that exhibit less risk and/or less cost compared to lower values.

Notably, the implementation context within which a project is implemented is exogenous to project delivery and cannot be changed or influenced by the manager or decision maker. Other elements of this delivery system can be considered as 'structural', in the sense that they formulate the structure of delivery. The business model (BM) and the governance arrangement are such elements. Defined and formulated at the front end, during the initiating stages of project implementation and the elaboration of the tendering process, they define the project's goals, organisation and overall setup. Over time, fewer changes may be made to improve performance. On the other hand, the Funding scheme and the Financing scheme are elements that may be modified (under specific conditions) throughout the life cycle of the project and induce incentives to various project stakeholders. For this reason, they may be used to promote strategies and support project-specific policies with respect to project delivery. Finally, the transport mode typology remains unchanged once the decision to deliver a specific infrastructure asset is taken. Considering the above differentiation in the way that they affect project delivery, system elements are grouped into: exogenous, structural, policy and transport mode typology.

Following this introductory section, Sections 3.2 to 3.5 present and discuss the development, validation and operationalisation of exogenous, structural, policy and transport mode indicators. Section 3.6 presents observations and remarks from the analyses conducted on the BENEFIT project case dataset after transforming case study information into indicators. The chapter ends with conclusions as per the usefulness of the indicators in both research/analysis and benchmarking.

3.2 Exogenous indicators: the implementation context as a pre-requisite for the delivery of transport infrastructure projects

Murwantara Soecipto, Tom Willems and Koen Verhoest

In general, the implementation context is considered important to the success or failure of a certain public policy. Empirical studies and, also, the qualitative analysis presented in Chapter 2 show the importance of the implementation context on the delivery of transport infrastructure projects. First, a stable and reliable institutional context is a vital aspect for creating a conducive investment climate and for achieving projects on time, within budget, and living up to their original objectives in terms of traffic and revenues. Second, favourable macro-economic and financial conditions are highly demanded by private actors to guarantee their investment (i.e. Hammami *et al.* 2006; Mahalingam and Kapur, 2009; Chan 2010; Galilea and Medda, 2010; Matos Castano, 2011; Delhi and Mahalingam, 2013; Zagosdzon, 2013; Percoco, 2014; Mota and Moreira, 2015).

Indeed, many policy experts point to 'context' as an important element for project success. As Peters and Pierre (2012) explained, the implementation context varies across countries and political systems, policy design and characteristics of the population.

With respect to the implementation context, there are indices provided by international organisations, describing various aspects of the implementation context. The objective is to review existing indices and, based on them, to formulate a comprehensive set of indicators that could provide a better understanding of the implementation context of transport infrastructure projects at country level.

Key criteria applied in the selection of relevant indices representing the implementation context were:

(i)   a measure of the political, regulatory, administrative, economic and financial conditions that are relevant for infrastructure project success;

(ii) as a minimum, available for the 18 countries of the BENEFIT project case dataset and preferably be available for many other countries either worldwide or within Europe;

(iii) a meaningful variation across the countries included;
(iv) data available over a long period (from 2000 or earlier till now) and with regular updates (e.g. yearly);
(v) widely used and accepted by the international community of practitioners/scientists;
(vi) no overlap, while at the same time giving a good and relevant picture of the involved dimension(s);
(vii) not mode – or project specific.

A combination of two different sets of indices developed by two different international organisations was chosen to get a deeper and richer insight into this complex notion of the two dimensions representing the implementation context: the first being the 'institutional context' and the second being the 'economic and financial context' (Vanelslander *et al.*, 2015, pp. 10–25). The combination of these two broad dimensions, results in a detailed and 'hands on'/usable set of indicators of how the implementation context looks like for transport infrastructure projects that are either publicly or privately financed. While the institutional context captures three dimensions: the political, regulatory and administrative setting, the economic and financial context delineates better understanding of the overall macro-economic and financial conditions of a certain country at a specific time of observation. The coming sub-sections discuss briefly these two broad dimensions of the implementation context.

3.2.1 Institutional context

This institutional dimension is crucial to delivering project success, as it contributes to the creation of a stable and reliable political, regulatory and administrative context for on-going projects from project preparation to implementation.

First, the 'political' sub-dimension 'political capacity, support and policies' is composed by three main governance indicators of the World Bank Governance Indicators (WGIs), that is political stability and absence of violence, control of corruption and voice and accountability. When combined, these three indicators give a good overview of the general political situation in a country.

(i) *Political stability (WGI)*, measures the perception of the likelihood of political instability and/or politically motivated violence, including terrorism. The basic political stability and the absence of violence are essential conditions for attracting private investment, as well as the delivery of transport infrastructure.
(ii) *Control of corruption (WGI)*, measures the perception of the extent to which public power is exercised for private gain, including both petty and grand forms of corruption, as well as 'capture' of the state by elites and private interests. Public infrastructure policies and projects are vulnerable to potential capture by private elites and interests.

(iii) *Voice and accountability (WGI)*, measures the citizen perception to participate in selecting government, freedom of expression and freedom of association and free media.

Second, the 'regulatory' sub-dimension 'legal and regulatory framework' is also composed of two governance indicators of the World Bank, namely, rule of law and regulatory quality, combined with the inverse of the aggregated OECD indicators of regulation in energy, transport and communications (ETCRs) on the regulatory restrictiveness of markets (i.e. measuring the level of liberalisation). Again, these three elements depict a rather comprehensive picture of the judicial and regulatory context of a country.

(i)   *Rule of law (WGI)*, measures the perception of the extent to which agents have confidence in and abide by the rules of society, and, in particular, the quality of contract enforcement, property rights, the police and the courts, as well as the likelihood of crime and violence. Effective rule of law is crucial when dealing with large and impactful public policy endeavours such as transport infrastructure projects. It may influence and guarantee private sector investment and its sustainability in the near future, as for example shareholders and stakeholders need to be able to rely on the judicial system to safeguard their basic (property) rights.
(ii) *Regulatory quality (WGI)*, measures the perceptions of the ability of the government to formulate and implement sound policies and regulations that permit and promote private sector development. Regulatory quality is a necessary component to build public trust, enhance effectiveness and efficiency in the public sector and to promote private sector involvement in transport infrastructure (and to keep them interested for more investments in the near future).
(iii) *Liberalisation of markets* regarding transport, energy and communications (the inverse of the ETCR developed by OECD), measures the degree of public ownership, vertical integration, entry regulation and market concentration.

Third, the 'administrative' sub-dimension 'public sector capacity' has only one indicator, namely government effectiveness, developed by WGI. It captures the perception of the quality of public services, the quality of the civil service and the degree of its independence from political pressures, the quality of policy formulation and implementation and the credibility of the government's commitment to such policies. In transport infrastructure projects, a strong and reliable public sector, with a sufficient level of expertise, experience and commitment, is crucial for building the necessary trust relationship with private sector companies.

The necessary data from the two sources (WGI and OECD-ECTR) were operationalised as a single indicator for the institutional context with values between 0 and 1. The method of aggregation for the institutional context index encompassed the following steps: (i) standardisation of raw data for all indicators at sub-dimension level between 0 and 1; (ii) calculation of an un-weighted average of indicators for each dimension; (iii) calculation of an un-weighted average of the three different dimensions for the institutional context.

Finally, the formulated InI was validated as per its reliability. A hierarchical cluster analysis was employed for the institutional index using the original data on the seven sub-dimensions and three years of observation, that is 2001, 2007 and 2013. The results show that the index reflects the underlying dimensions over time, thus the use of institutional index is defendable as it reflects sufficiently the underlying dimensions. In addition, using a Spearman rank correlation, it was also found that the institutional index is robust (see Roumboutsos *et al*, 2016b, pp. 73, 318–330).

3.2.2 Financial-economic context

Two aspects were considered for the better understanding of the economic and financial conditions of a certain country.

First, the overall '(growth) competitiveness index (GCI)' of the World Economic Forum (WEF) aims at measuring the capacity of the national economy to achieve sustained economic growth over the medium term, controlling for the current level of economic development. It is basically an index about future economic growth and investment potential. This index is substantially different from the governance indicators of the World Bank. Originally, only the 'macro-economic' dimension of the index (with a reduced potential overlap with the institutional context index) was considered, but this had two important disadvantages: it is only available from 2001 onwards, and does not include important and relevant information on technology, innovation, etc. of a certain country.

Thus, despite its name, the FEI measures more than just the macro-economic and macro-financial context in a country. More broadly, it reflects the *business environment* in a country and may be considered as a proxy for a country's *level of productivity*. It includes a macro-economic dimension (capturing the government budget balance, gross national savings, inflation, general government debt and the country's credit rating), as well as a financial market development dimension (measuring among others the availability and affordability of financial services, ease of access to loans, soundness of banks and venture capital availability). In addition, this index also includes information on supporting contextual elements and policies, such as the goods market efficiency, labour market efficiency, technological readiness, market size, business sophistication and innovation in a country. Moreover, the index captures the availability of some basic requirements in terms of education, health of the population and overall infrastructure, as well as some limited business-oriented aspects of the institutional environment (e.g. property rights, intellectual property protection, efficiency of legal framework in settling disputes, strength of auditing and reporting standards). It should be noted that there is some limited overlap with the InI defined previously, particularly in the sub-dimension of the legal and regulatory framework, as the 'growth competitiveness index (GCI)' captures some institutional aspects with direct bearing on the business environment in a country.

Second, as supplement to the index, four key indicators regarding the macro-economic conditions such as inflation (consumer prices), general government final consumption expenditure, GDP per capita growth and unemployment rate and, also, the two key financial indicators such as S&P global equity prices and domestic credit to private sector are introduced. These key indicators are important as background information when considering the implementation context of a project in a country, but cannot easily be rescaled to a simple score or grade. They serve as additional data for the aforementioned macro-economic financial context score. They are complementary in nature, and, when combined, give a complete overview of the financial economic implementation context.

3.2.3 The evolution of implementation context for 18 European countries

The implementation context for the 18 European countries of the BENEFIT project case dataset was mapped using the indicators described previously. The InI varies across countries, with a highest average score for Finland (0.86) and lowest for Serbia (0.43). The average score across the 18 countries between 2001 and 2014 is 0.71. Figure 3.1 depicts the evolution of the institutional context index for the 18 countries over this period.

Figure 3.1 shows that the institutional indicator is rather stable over time and remains largely unaffected by the GFC. In effect, when comparing the change of institutional context indicator from 2001 to 2014, the results show an average annual increase of 0.4 per cent. However, this result does not reflect the full variation in the sample. As an example, coming from southern European countries, Greece experienced a deterioration of its institutional context over that period, with its indicator dropping by 0.5 per cent per year on average. Sweden experienced a stable institutional indicator over the same period.

The FEI is represented by the 'GCI' from the WEF. This index indicator also varies quite strongly across countries, with a highest average value for Norway (0.78) and a lowest value for Serbia (0.44) between 2001 and 2014. The average score for this period across the 18 countries is 0.61, as illustrated in Figure 3.2.

As visible from Figure 3.2, the FEI is much more variable over time, compared to the InI. All 18 countries experienced an 'economic shock' during the GFC. By considering 2008 as a benchmark, the FEI dropped significantly afterwards with a reduction of 8 per cent on average. The United Kingdom had the largest decrease (roughly 15 per cent), while Norway the smallest (about 1 per cent). The results also show the average FEI having a small annual increase of about 0.1 per cent from 2001 to 2014. However, while some countries were able to show a full recovery, such as Norway, Germany, the Czech Republic or Poland, others, such as Greece, Portugal, Croatia, Slovenia, Spain or Italy, continued to decline, with Greece showing a small recovery only in 2013.

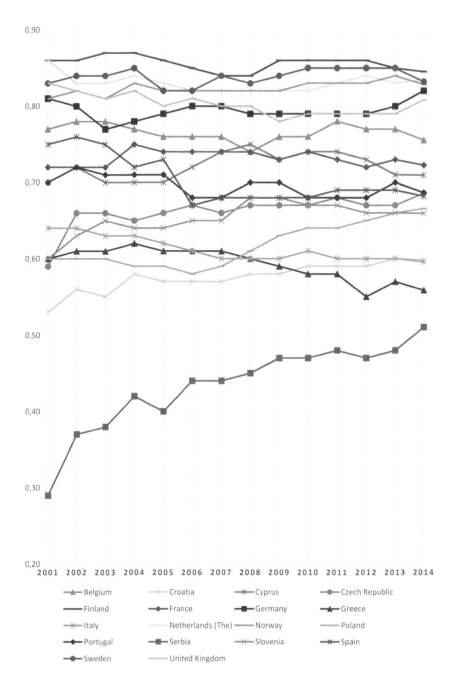

Figure 3.1 Institutional Indicator (InI) across 18 European countries over the period 2001–2014

Source: Authors

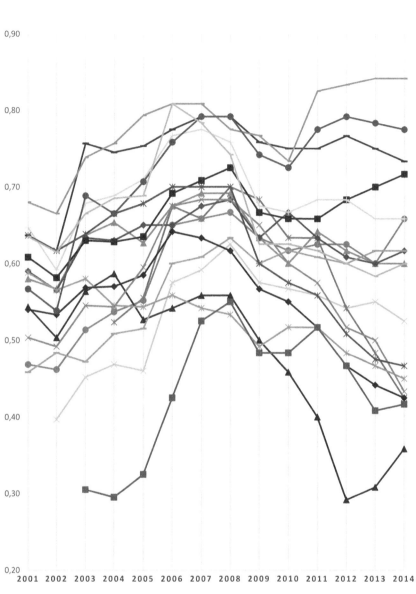

Figure 3.2 Financial-Economic Indicator (FEI) across 18 European countries 2001–2014

3.3 System endogenous structural indicators

Ibsen Chivata Cardenas, Hans Voordijk and Athena Roumboutsos

Meeting project management goals, usually reflected by achieving cost and time to completion targets and performance-aligned owners' strategic goals, has been highlighted by various studies (Cooke-Davies, 2004; Shenhar and Dvir, 2007). Project governance (PG) is, in principal, related to the achievement of project management goals, while the implemented BM sets the scene for achieving owners' goals. Both elements are part of the transport infrastructure delivery and operation/maintenance system (BENEFIT Framework) describing its structure in terms of its value proposition and how it is meant to address user needs (expressing owners' goals). Both also reflect how the contractual setup is organised to deliver the infrastructure asset and its services.

Finally, both elements are endogenous to the project, that is they can be influenced by project stakeholders. However, once set, usually at the front-end of the project, the ability to change them diminishes dramatically as time elapses. In this context, governance and the BM are considered 'structural' elements of project delivery.

The present section describes the theoretical background, the validation process and the operationalisation of indicators for these two structural elements.

3.3.1 Business models for transport infrastructure

Athena Roumboutsos

3.3.1.1 Business models

BMs are a new unit of analysis and represent a system-level concept (Zott *et al.*, 2011). The subject has received considerable contributions, especially since 1995 (Burkhart *et al.*, 2011; Zott *et al.*, 2011), and as a new concept many definitions, classifications and attributes can be found in the literature. Zott *et al.* (2011), in their review, grouped available definitions as: a statement (Stewart and Zhao, 2000), a description (Weill and Vitale, 2001), a representation (Morris *et al.*, 2005; Shafer *et al.*, 2005), an architecture (Dubosson-Torbay *et al.*, 2002; Timmers, 1998), a conceptual tool or model (Osterwalder, 2004) and a structural template (Amit and Zott, 2001). However, several common features are identified in all previous representations of a BM (Burkhart *et al.*, 2011): (i) BMs are applicable across all business types; (ii) BMs include static and dynamic aspects and (iii) a component-based perspective is endorsed in all definitions. Components are understood as interdependent units. In the present analysis, the BM refers to components or 'building blocks' that produce a proposition that can generate value for consumers and, thus, for the organisation (Demil and Lecocq, 2010).

The BM concept may be viewed as 'static' (i.e. as a blueprint), suggesting best practice, or 'dynamic', taking into consideration 'change' (Demil and Lecocq, 2010). The latter enforces the concept of a system's perspective (Zott and Amit, 2010) and its key design themes: novelty, lock-in, complementarities and efficiency (termed *NICE*), as well as its relation to the Resource-Based View (RBV) (George and Bock, 2010). Hamel (1999) suggests that firms must acquire resources concomitantly to the implementation of new BMs. Notably, a BM is related to strategy, value creation and value capture (Zott and Amit, 2011) and may be employed as an 'opportunity' facilitator (George and Bock, 2010).

In a project setting, 'value' concerns 'use-value' (Hjelmbrekke *et al.*, 2017). Bowman and Ambrosini (2000) refer to specific qualities of the product (e.g. infrastructure) as perceived by customers (users) in relation to their needs. Value capture is then related to the potential of use-value to generate strategic value, as well as the resources needed and their organisation (RBV) in order to serve strategic objectives. Simultaneously, a project retains its key project management objectives such as meeting cost and time targets. In this context, Watson's (2005) BM definition is more appropriate: a BM describes a company's operations, including all its components, functions and processes, which result in costs for itself and value for the customer. Therefore, the ambition is to achieve low cost and high value resulting in benefits (including financial returns). A successful BM depends on the ability of an organisation to introduce 'unique capabilities or competencies that have a special value to the customer base' (Reading, 2002).

The above paragraphs formulate the context within which the indicators describing the transport infrastructure BM have been developed. Notably, two distinct indicators have been introduced reflecting the dual ambition of the transport infrastructure BM: (i) the Cost Saving Indicator (CSI) and (ii) the Revenue Support Indicator (RSI; Vanelslander *et al.*, 2015, pp. 38–47).

Finally, while the relation and impact of the BM on performance has been recognised, there is insufficient knowledge regarding component interdependencies within the BM and between them, as well as limited insights on criteria and metrics for an appropriate evaluation of BM (Burkhart *et al.*, 2011). To this end, this work presents a significant contribution to BM research.

3.3.1.2 *Business model indicators for transport infrastructure delivery*

Key inspiration in the formulation of the BM indicators for transport infrastructure delivery has been drawn by the work of Zott and Amit (2010) on the system's perspective of the BM and the design key themes (i.e. novelty, lock-in, complementarities and efficiency, coupled with notions of the RBV in BM by George and Bock (2010).

Based on evidence reported in the literature, four dimensions are identified, which are descriptive of the robustness of the BM with respect to securing revenue streams and reducing costs (Vanelslander *et al.*, 2015, pp. 39–42; Mladenovic *et al.*, 2016, pp. 118–123; Roumboutsos *et al.*, 2016b, pp. 75–96):

DIMENSION 1. REVENUE ENHANCING THROUGH 'BUNDLING'

Within this dimension, (described as 'complementarity' by Zott and Amit, 2010), two types of bundling were identified. Bundling with:

- transport-related services and infrastructure with a scope of reducing the revenue/demand risk. These concern:
  - investments in infrastructure, which are additional to the principle (or core) investment and may be assessed as the ratio of the core investment over the total investment;
  - brownfield transport infrastructure with known and secure demand. In this case, the assessment is made through the ratio of known or demonstrated demand over the total demand forecast.
- non-transport related activities/services.

DIMENSION 2. REVENUE PROTECTION

Revenue protection corresponds to a large degree to 'lock-in' as described by Zott and Amit (2010) and concerns:

- protection against competition (positioning of infrastructure in the network and/or induced through the contract) in combination with required coopetition as foreseen in the planning of the transport network. Notably, the position of the infrastructure in the network is characterised by its level of exclusivity (monopoly status) and level of integration with the transport network. A favourable combination, either natural or contractually induced, may protect the project's revenue stream(s);
- provision of services creating 'captive' users.

DIMENSION 3. COST EFFICIENCY (SAVING) OF CONSTRUCTION, OPERATION AND MAINTENANCE

This dimension corresponds to 'efficiency' described by Zott and Amit (2010) and, also, 'novelty' as innovation is adopted to reduce the cost of construction and/or the cost of operation and maintenance. Within transport infrastructure delivery, this dimension is characterised by:

- life cycle planning;
- the adoption of innovation or other efficiency interventions in construction and/or operation;
- optimal risk allocation with respect to risk connected to the BM (design, construction, operation, maintenance, exploitation and others).

DIMENSION 4. AGENTS' CAPABILITY TO MANAGE THE BM

This dimension connects the BM with the Resource-Based View (RBV) (George and Bock, 2010) and refers to the competence required to implement the BM. By 'agent', each actor responsible for implementing the various activities and components of the BM is referred to. In addition, in a project context, it refers to assessing the proper allocation of the various project risks with respect to the competence of the actors to manage them. Finally, in a transport infrastructure

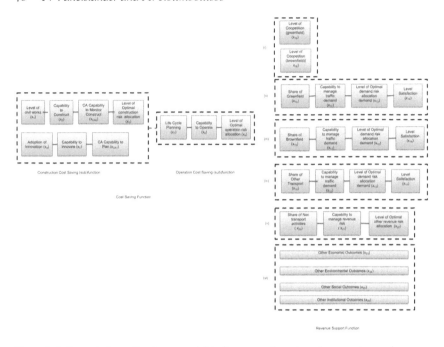

Figure 3.3 Formulation of business model indicators as functions (cost saving and revenue support)

Source: Author.

BM, the competence of the contracting authority, that is the ultimate 'owner', is also of essence.

Considering a system's perspective (Zott and Amit, 2010), the configuration of the above dimensions as functions of the cost saving and revenue support are illustrated in Figure 3.3. The system representation follows two basic rules: interdependent factors are structured in a series configuration, while independently contributing factors in a parallel configuration. Therefore, in the formulation of the two indicators, the product of the factors in series is summed with the respective ones in parallel. All factors and resulting indicators are unitised.

3.3.1.3 *Operationalisation and validation of the variables measuring the business model*

Several criteria were considered for the operationalisation of the factors identified to be included in the Cost Saving and Revenue Support Indicators. A key criterion was their ability to be readily and accurately assessed by a wider professional audience. To this end, information from case studies, which was in the public domain was only used. For some factors, a binary (yes-no) assessment was made, while in others, a qualitative scale was employed guided by assessment questions (see BENEFIT project reports Vanelslander *et al.*, 2015, pp. 38–47; Mladenović *et al.*, 2016, pp. 118–123; Roumboutsos *et al.*, 2016b, pp. 92–96).

Validation was conducted on two levels: for individual factors and for the product of factors in a series configuration (see Figure 3.3). In both cases, the impact on project outcomes (cost and time to completion, actual traffic and revenues versus forecast) was examined through quantitative and semi-quantitative approaches based on hypothesis testing. With respect to individual factor testing, semi-quantitative analysis consisted of reviewing all BENEFIT project case dataset with respect to testing certain hypotheses using values from indexes. The full quantitative analysis consisted of various methods: Non-parametric statistical testing of correlations, sensitivity analysis (see next section) and Principal Component Analysis. In these analyses, additional factors not originally considered for inclusion in the BM indicators, were also tested. Table 3.1 presents key factors tested along with the respective hypotheses and findings.

The product of the factors in a series configuration (see Figure 3.3) was also tested for its impact on performance outcomes. More specifically:

- Non-parametric statistical tests were used to test for correlations.
- Sensitivity analysis and Bayesian Networks were employed to test for the relative contribution of each factor in the final outcomes.
- principal component analysis was used to identify whether some factors could be dropped from the composite indicators or if indicators should be assigned.

All the above analyses concluded that the identified indices and their products, when in a series configuration (see Figure 3.3), were important. The principal component analysis did not justify any relative weighting of components. Therefore, the final configuration of the BM indicators is as follows:

COST SAVING FUNCTION – INDICATOR

The CSI is formulated as follows:

During the construction phase: $\Phi_{CS} = x_2 x_3 x_{CA2} - (x_1 - x_1 x_2 x_3) + x_4 x_5 x_{CA1} + x_7 x_8 x_9 x_{CA3}$

During the operation phase: $\Phi_{CS} = x_7 x_8 x_9 x_{CA}$

When $\Phi_{CS} = 0$, no cost saving is observed. When $\Phi_{CS} < 0$, then cost overruns or a tendency for cost increase may be observed. Both expressions of Φ_{CS} are normalised so as $\Phi_{CS} \leq 1$.

REVENUE SUPPORT FUNCTION

The RSI is formulated as follows:

$\Phi_{RS} = (x_{11} + x_{16}) * \%$ of brownfield $+ x_{10} x_{12} x_{13} x_{14} + x_{15} x_{17} x_{18} x_{19} + x_6 x_{12} x_{13} x_{14} + x_{20} x_{21} x_{22}$

Table 3.1 Business model hypotheses tested

Factor	Combined with	Hypotheses		Findings
Level of Control	Risk allocation	H1:	Appropriate risk allocation based on level of control leads to improved performance with respect to traffic (even under adverse macroeconomic conditions)	TRUE
Capability of the operator	Level of control	H2:	The capability of the operator to manage demand risk and more so the supply chain contributes to project performance	TRUE
		H3:	The capability of the operator contributes in achieving traffic targets	TRUE
Capability to construct	Level of construction difficulty	H4:	The capability of the constructor contributes in achieving construction cost and time targets	Inconclusive
Capability of the contracting authority		H5:	The capability of the contracting authority to plan in achieving performance targets	TRUE
		H6:	The capability of the contracting authority to monitor project implementation in achieving cost and time targets	TRUE
		H7:	The capability of the contracting authority to manage stakeholders in achieving cost and time targets	TRUE
Bundling of activities	Capability of the operator	H8:	Integrating alternative revenue activities (e.g. Commercial activities) and combined transport infrastructure delivery contributes to project performance	TRUE
Investment size		H9:	Achieving construction targets is dependent on the size of the investment	NOT TRUE
Infrastructure configuration		H10:	Achieving construction targets is dependent on infrastructure configuration (node – link)	NOT TRUE
User type		H11:	Achieving construction targets is dependent on user mix	NOT TRUE

Source: Author.

The component (v) (see Figure 3.3) referring to general outcomes *has been dropped* providing a more 'operational' configuration of the indicator.

3.3.2 Governance in infrastructure contractual arrangements

Ibsen Chivata Cardenas and Hans Voordijk

3.3.2.1 Governance

Governance is a decision framework, which is critical in setting direction, monitoring performance and responding to external pressures, and has been studied through the lenses of (Clarke 2007; Mallin 2004, 2006):

- agency theory, as the firm is viewed as a set of contractual relationships between the owners as 'principals' and the directors of the firm as their 'agents';
- transaction cost economics that incorporates the notion of a series of contracts among various players to overcome the limitations of a single contract between the agent and the principal. The set of contracts is a governance structure that corrects any misaligned actions;
- stewardship theory, by which optimum governance structures can nullify the inherent conflict of interest between owners and managers. Accordingly, company directors are regarded as stewards of the company's assets who will act in the best interests of the shareholders. Stewardship theory is informed by theories of motivation, power and situational factors such as management philosophy and culture;
- stakeholder theory, whereby equal attention is paid to internal and external stakeholders, with corporate social responsibility extending this scope;
- institutional theory, where governance is described by relations, norms and policymaking. Here, policymaking is largely seen as the result of interactions between a multitude of actors (Conteh, 2013). According to this perspective, traditional hierarchical management styles have been replaced by hybrid multi-actor and multi-level governance along with mutually interdependent decision-making structures (de Bruijn and ten Heuvelhof, 2008): the major issue in this approach is the alignment of interests in defining the overarching scope to be achieved.

In terms of governance in organisations, one can distinguish two types of environments. For a government, the definition of goals serving public interest is the major governance aim under scrutiny. Meanwhile, for a private company, governance is defined as 'a set of relationships between a company's management, its board, its shareholders and other stakeholders. This type of governance, namely corporate governance, also provides the structure through which the objectives of the company are set, and the means of attaining those objectives and monitoring performance are determined' (OECD 2004, p. 11; Klakegg *et al.*,

2008). For a project, governance is considered as a framework for decision making, encompassing the process of making decisions and the established framework, models or structure for their enablement (Garland, 2009). In the construction of this decision framework, policymaking becomes an inter-organisational process in which policies need to be conceived as the result of an interaction between several stakeholders (Conteh, 2013). The assessment of multiple relations, incentives and objectives takes the complexity of the decision framework to a much higher level. Thus, in order to respect the social principles of fairness, equality and equity, governance rises as an institutional environment for the development of collective decision-making in settings where there is a plurality of actors or organisations and where no formal control system can dictate the terms of the relationship among these actors and organisations (Chhotray and Stoker, 2010).

Most of the work in the categories proposed by Chhotray and Stoker (2010) comes from the analytical works of Williamson (2000) which concerns transaction costs in governance structures and the economic agency theory. In this perspective, the governance rules adopted, the level of collectivity, the decision-making process and the level of decentralisation will also determine the level of transaction costs involved.

Notably, transport infrastructure bears the characteristics of a 'public good' and, therefore, the interests considered when addressing decisions with respect to its delivery and operation are multiple. Moreover, the delivery of infrastructure projects is the common ground of public sector governance, corporate governance and PG, where goals and incentives need to be aligned.

While considerable research has focused on methods by which projects deal with unique, novel and transient operations, PG did not receive much attention until rather recently (Abednego and Ogunlana, 2006; Crawford and Cook-Davies, 2005). As a mode of organising transactions, PG presents a multi-dimensional phenomenon, encompassing the initiation, termination and maintenance of the on-going relationships between a set of parties (Heide, 1994). Moreover, PG is project-focused and describes how the project management processes are governed throughout the project life cycle (White, 2001; Winch, 2001).

In general, in the context of projects, governance could be thought of as building consensus necessary to achieve an objective in an arena where many different interests are at play. In the project environment, governance mechanisms are needed to support the operational control processes, and to manage the interface between project teams and their clients (Turner and Keegan, 2001).

Furthermore, in the context of PPPs, Guo *et al.* (2013) and Reeves (2013) define governance as the phenomenon of steering and coordinating a PPP by setting up organisational structures, running decision-making procedures and using instruments such as contracts and agreements that do not rest solely on the authority and sanctions of government. First, structure concerns the actors and institutions involved in a PPP (Scharpf, 1991). Then, procedure brings dynamics in and covers PPP decision-making procedures as they run from the initiation phase to the operational phase (Verhoest *et al.*, 2012). The procedural approach seems to be better positioned than the structural one to describe the actual relationship between actors over time. Procedures can be defined contractually or not for the implementation and the operation phase of a project, respectively. Finally, instruments are the tools used by the government to steer a PPP towards the achievement of its objectives (Verhoest *et al.*, 2004). More specifically, in any project, the structure is defined by the procurement process.

It should be noted that in transport infrastructure, governance could be analysed at the project level. This entails focusing on the relationships between a contracting authority and a contractor or contractors. These relationships are usually forged by transactions during a procurement process and are tangibly reflected in a contract and its changes. These features also indicate that the above-mentioned theoretical lenses of agency theory and transaction cost economics appear to be suitable approaches to understand governance in transport projects, and this is the perspective advocated in this book. The next section further reviews recent research studying PG in infrastructure projects focused on transactions and describing relationships between the client and contractors.

3.3.2.2 *Governance in infrastructure contractual arrangements*

Based on the background provided on governance in the previous section, it is evident that PG is connected to the characteristics of the relations between the contracting authority and the contractor or contractors (Voordijk *et al.*, 2015).

Recently, important theoretical developments have been reported in infrastructure governance. For instance, Li *et al.* (2012) and Chen and Manley (2014) have developed and extensively tested conceptual models in which relevant PG instruments and factors are identified and related to both project performance (PP) and the reduction of transaction costs in construction projects. Chen and Manley's (2014) research focused on collaborative infrastructure projects, while Li *et al.* (2012) researched the influence of complexity and uncertainty factors on the occurrence of transaction costs in projects. In so doing, the latter authors identified specific methods to cope with the uncertainty of a project to ultimately reduce transaction costs. These authors show that: (i) integrated design and construction, (ii) early contractor input into the design, (iii) a healthy competition between bidders, (iv) a fair allocation of risks through bonding and contract clauses that establish incentives/disincentives and (v) a design that is sufficiently complete, are approaches that assist in coping with uncertainty in construction project transactions, and should ultimately reduce transaction costs.

Meanwhile, Chen and Manley (2014) found that the influence of formal governance mechanisms on PP is mediated by an informal governance mechanism. Formal governance is seen as comprising contractual incentives for clear and equitable risk allocation, whereas informal governance comprises non-contractual incentives to enhance mutual trust, enable cooperation, facilitate open communication and share knowledge. More specifically, Chen and Manley (2014) identified essential mechanisms that define formal and informal governance for collaborative infrastructure projects. Three mechanisms can define formal governance: (i) collective cost estimation, (ii) risk and reward sharing regime and risk sharing with service providers and (iii) design integration. Four social-based mechanisms define informal governance: (iv) leadership, (v) relationship management, (vi) team workshops and (vii) communication systems.

Susarla (2012) has also shown the importance of mechanisms in enhancing governance in information technology service contracts such as contractual flexibility and actions that enable contracting authorities to maintain bargaining power. Evidence of the relevance of these issues in transport projects was gathered and analysed (Roumboutsos *et al.*, 2016).

Table 3.2 summarises the main aspects of the comprehensive reviews conducted by Li *et al.* (2012), Susarla (2012) and Chen and Manley (2014), which were the basis of the conceptual models hypothesised and tested along with the hypothesised effects of some variables linked to PG.

It could be argued that most of the informal features of governance, namely (iv) leadership, (v) relationship management, (vi) team workshops and (vii) communication systems are inherently embedded in the formal ones considered in Table 3.2. For instance, actions to prevent a contractor's opportunistic behaviour, which are typically related to informal mechanisms of governance (Susarla, 2012), are embedded in the 'flexibility of the contract' and the 'contracting authority's bargaining power' (unilateral termination) formal governance mechanism variables. Notably, opportunistic behaviour is typically reflected in contractor's actions such as reducing the quality of work, lodging claims, disputes or litigation, refusing to share information or hindering problem solving (Zhang *et al.*, 2016), and are opposing behaviours to those stimulated by informal mechanisms of governance. In addition, the literature is not conclusive on the relevance of informal governance to PP and, therefore, including the informal governance variables [(iv), (v), (vi) and (vii)] may ultimately be problematic (Dyer and Singh, 1998; Knights *et al.*, 2001; Luo, 2002; Poppo and Zenger, 2002; Li, Arditi and Wang, 2012; and Chen and Manley, 2014).

Based on the above review, Table 3.3 compares identified specific variables with those considered herewith to measure governance in transport infrastructure projects together with their suitability for endorsement. These variables underwent further checks to examine disjointedness, subtlety and measurability. The evaluation of the variables' consistency with the information in the BENEFIT project case dataset was also carried out.

3.3.2.3 *Operationalisation and validation of the variables measuring governance*

Based on the quantitative research by Chen and Manley (2014), Li, Arditi and Wang (2012) and Susarla (2012), a set of relevant variables measuring PG were identified. These have been extensively validated with statistical tests and exhibit high factor loadings. Factor loadings reveal the extent to which a certain variable

Table 3.2 Hypothesised effects of some variables on aspects of project governance

Variables	(a) Hypothesized effects on aspects of project governance
Collective cost estimation (G1, G2)*	reduce uncertainty in the project; facilitate contractors' contributions in the design stage which may establish a trust-based cooperative relationship; provide complete plans and specifications which may decrease the number of disagreements and disputes; maximize outcomes, since the owner benefits from a range of design scenarios, sensible risk management (risk awareness), and appropriate contract development.
Competition between bidders (G3)**	provide incentive to achieve cost-effective pricing; engage optimal technical competence and experience, i.e., appropriate qualifications and experience, is conductive to speedy decisions, smooth operations, few reworks, and easy communication, all of which contribute to lower transaction costs; engage competent contractors who may operate efficiently and may promote a problem-free environment, hence contributing to a more stable environment with lower transaction costs.
Integration design and construction (G4)**	improved integration, collaboration and communication in the interface between design and construction may reduce uncertainty; facilitating contractors' contributions in the design stage may establish a trust-based cooperative relationship; provide complete plans and specifications which may decrease the number of disagreements and disputes; for maximizing outcomes, since the owner benefits from a range of design scenarios, sensible risk management (risk awareness), and appropriate contract development. increases financial certainty for contractors, thereby reducing the likelihood of margin recovery strategies such as claims.
Disincentives (G5, G6)(*), (**)	discourage opportunistic behavior on the part of the contractor; motivate the contractor to minimize project duration.
Profit allocation (G7)*	affect the collaborative behavior of participants involved in collaborative projects.
Bonding requirements (G8)(*),(**)	discourage opportunistic behavior on the part of the contractor; motivate the contractor to minimize project duration.
Allocation of risk (G9) (*),(**)	a 'fair' allocation of risks between the parties may reduce transaction costs.
Flexibility of the contract (G10)***	these flexibility provisions reduce the likelihood of rent seeking by both parties lowering maladaptation and underinvestment.
Termination without cause arrangement (G11)***	grants unilateral control to the client, reducing the likelihood that a contractor can engage in opportunistic rent seeking. The threat of unilateral termination by the client correspondingly lowers the likelihood of

continued . . .

Table 3.2 Continued

Variables	(a) Hypothesized effects on aspects of project governance
	underinvestment by the contractor. Such enhanced performance incentives for a contractor correspondingly lower incentives for a client to force concessions or strategic termination, lowering the likelihood of maladaptation and underinvestment, facilitating smooth adaptation to unfolding contingencies.

*   Variables found to be statistical significant factors of PG. Note between PP and PG was found also a significant positive association (Chen and Manley, 2014).

**   Variables found to be statistical significant factors that ultimately contribute to reduce transaction costs (Li *et al.*, 2013)

***   Variables found to be associated significantly with Pareto improving amendments. Pareto improving amendments are considered as renegotiations that improve the welfare of one party without worsening the other (Susarla, 2012).

Source: Authors

is relevant to measure PG. These were further validated with respect to their relation to PP. To this end, the following set of hypotheses (some of which had been proposed earlier by Eriksson and Westerberg (2011)) were formulated:

H_1: The higher the level of integration between client and contractors *in the design stage* (variables G3 and G4), the better the PP in terms of (i) *cost and* (ii) *time to construction completion.*

H_2: The higher the level of integration between client and contractors in the pricing stage (variables G1 and G2), the better the PP in terms of (i) *cost and* (ii) *time to construction completion.*

H_3: The more the payment is based on incentives/disincentives (variables G5, G6, G7, G8), the better the PP in terms of *(a) Cost and (b) Time to construction completion.*

H_4: The existence of contractual clauses that enable both updating of service and price changes (flexibility contract provisions, variable G10) are associated with greater PP in terms of (i) *cost and* (ii) *time to construction completion.*

H_5: The existence of contractual termination clauses (prematurely without cause, variable G11) in a contract are associated with greater PP in terms of (i) *cost and* (ii) *time to construction completion.*

H_6: The concentration of revenue risks on one contractual party (variable G9) is negatively associated with PP in terms of (i) *cost and* (ii) *time to construction completion.*

Based on the information available in the BENEFIT project case dataset for the eleven (11) variables, the coding presented in Table 3.2 was applied.

The analysis framework for the validation was based on a sensitivity analysis and Bayesian Networks. Sensitivity analysis is the study of how the uncertainty in the output of a mathematical model or system (whether numerical or otherwise)

Table 3.3 Variables endorsed from the literature review and their respective coding in BENEFIT project

Variables	(a) Variables identified literature and considered in BENEFIT* (endorsed/ not endorsed)	(b) BENEFIT Governance Indicator Index coding
Collective cost estimation (G1, G2)	(considered/ endorsed)	Coded as 0.5 if the contracting authority selected only one bidder to participate in the pricing stage and 1 if several bidders were involved in the pricing process. In the model this variable can take values as follows: dom(Collective cost estimation)={More than one bidder, One Bidder, None}
Competition between bidders (G3)	(considered/ endorsed)	This variable is coded as 1 (otherwise 0) for procurement processes in which more than one bidder was involved. In the model this variable can take values as follows: dom(Competition between bidders)={Yes, No}
Integration design and construction (G4)	(considered/ endorsed)	This variable was coded as 1 (otherwise 0) for contracts in which design and construction of works are combined within a single contract. In the model this variable can take values as follows: dom(Integration design and construction)= {Yes, No}
Disincentives (G5, G6)	(considered/ endorsed)	This variable takes a value from a three-point scale depending on the presence of two contract clauses. The variable is coded as 0.5 if the contractor solely carries the risk of rising costs and 1 if the contractor is additionally obliged to pay a penalty if completion dates are not met. If none of these conditions applies, this variable takes the value 0. In the model this variable takes a value as follows: dom(Disincentives)={Contractor pays a penalty plus carries the risk of rising costs, contractor only carries the risk of rising costs, None}
Profit allocation (G7)	(considered/ not endorsed, the BENEFIT data did not corroborate its relevance)	This variable is to be coded as 1 (0 otherwise) if profit due to cost underruns is to be shared into equal proportions.
Bonding requirements (G8)	(considered/ endorsed)	This variable was coded as 0.5 if clauses indicate that guarantees over performance were agreed (coded as 0 if no such clause). Coded as 1 when some additional incentives or disincentives are given to the contractor to increase performance (during construction or operation phases). Changes to sources and shares of remuneration streams (typically reflected in changes towards adoption or strengthening of availability and performance-based

continued . . .

Table 3.3 Continued

Variables	(a) Variables identified literature and considered in BENEFIT* (endorsed/ not endorsed)	(b) BENEFIT Governance Indicator Index coding
		remuneration streams) justify such a coding. In the model this variable takes a value as follows: dom(Bonding requirements)= {additional incentives or disincentives are given during execution, guarantees of performance agreed on, None}
Allocation of risk (G9)	(considered/ endorsed only for revenue risk since the other type of risks did not show significance)	This variable was coded as 0 (otherwise 1) if revenue risk allocation is concentrated in one party. In the model this variable can take values as follows: dom(Allocation of revenue risks)= {risk are mostly shared, risks are mostly concentrated}
Flexibility of the contract (G10)	(considered/ endorsed)	This measure was coded as a binary scale depending on the presence of clauses that specify updating of service terms or enable price changes. The measure is coded as 1 if either or both of the clauses are present. In the model this variable can take values as follows: dom(Flexibility of the contract)={Yes, No}
Termination without cause arrangement (G11)	(considered/ endorsed)	This variable was coded as 1 (otherwise 0) if clauses indicate that the contracting authority has an option to prematurely terminate the agreement without cause. In the model this variable can take values as follows: dom(Termination without cause arrangement)={Yes, No}

Source: Authors.

can be apportioned to the various sources of uncertainty in its inputs (Saltelli *et al.*, 2008). Sensitivity analysis is an ideal method to evaluate models (Borgonovo and Plischke, 2016). Bayesian Networks (BNs) are essentially a tool for modelling the relationships between variables using conditional probabilities (van der Gaag, 1996). These conditional probabilities can be learnt under certain conditions from small datasets, as shown by Onisko *et al.* (2001). The developed Bayesian Network models were evaluated using traditional tests reported by Anderson *et al.* (2004) and Lee and Moore (2014). Such tests include independence tests to check marginal and conditional independence among factors in the models, marginal log-likelihood estimation, which is a comparative measure and it is used to assess the goodness of fit, and cross-validation to verify the capability prediction of the models developed. In this analysis, the relationships of variables were confirmed by the rejection of the hypothesis of independence for each marginal and conditional relationship (using the cut-offs $p < 0.001$, $p < 0.01$ and $p < 0.05$ for different relationships). In this process, information from 40 European projects with data reflected in more than 50 records was used (see Appendix). As a result of this analysis, the indicator G7 in Table 3.2 was identified as irrelevant and accordingly removed from the set of indicators initially selected (Roumboutsos et al., 2016b, pp. 97–99; Mladenovic et al., 2016, pp. 123–126).

3.4 System endogenous policy indicators

Aristeidis Pantelias

Financing enables the implementation of transport infrastructure projects. It consists of raising capital at the front end of a project to construct and develop it. Raising this capital involves various providers who commit their funds in expectation of certain returns. Financing can only be committed if sufficient funding streams have been identified and secured that will be able to repay the initial capital raised.

Funding schemes refer to both the origin of revenue streams, that is by whom and how the revenues are provided, as well as to the method of remuneration, that is to whom and how the revenues are attributed. The two (i.e. revenues and remuneration) may coincide or may differ.

To successfully finance a transport infrastructure project, it is necessary to understand how it is developed. Different development phases represent different financial circumstances for the project, which need to be taken into consideration when evaluating its viability and determining the most appropriate financing scheme for it.

Almost every transport infrastructure project starts with a construction phase, where the infrastructure is built or renewed. During this phase net returns (or net cash flows) are negative. After that, it enters an operation phase, where the infrastructure asset is commissioned and used. During that phase, net returns (are expected to) turn positive. This holds particularly true for greenfield projects that start with a, sometimes, long construction phase (e.g. building a new port, airport, road or railway line) before entering a very long operation phase. For brownfield projects that start with mature assets already in place, the construction phase may be smaller in terms of both cost and time, but there is still a starting investment needed (for renewing, rehabilitating or extending the existing infrastructure). For these assets and during this initial investment phase, net returns will also be negative for some time not least since construction works usually hinder their (full) utilisation. In both cases, the operation phase includes a 'maturing' phase during which the project reaches its nominal traffic volumes. This is a very crucial and vulnerable point in a project's life cycle.

For infrastructure projects to be successfully financed and implemented, the following viability condition needs to hold: the positive net returns from the operation phase must be large enough to repay for the deficits made during the construction phase. This implies, first, that the financial arrangements made at the front end of the project must bridge the time-gap between the expenses made at the beginning and the cash inflow coming thereafter. Second, it is necessary that the net returns of the operating phase (which are directly linked to the underlying funding schemes) are large enough so that the financial arrangements can be fulfilled. Due to the large amounts of money and the extended time horizon of cash in- and out- flows involved (particularly in the case of greenfield projects), the financial decisions of the investors, that is whether and at what conditions to provide finance, contain substantial elements of 'speculation' or dealing with uncertainties, especially with respect to the risks associated with the revenue streams. Consequently, all risks and opportunities that may affect the project during its life cycle in any economically significant way are relevant for the financial decisions and will be reflected in the structure and cost of the financing. Therefore, understanding the various risks of an infrastructure project is of paramount importance for the determination of its financing scheme. Risks are analysed extensively in every project during its structuring at the front end and risk management is a process that goes on throughout the life cycle of a project.

Considering the above, it is expected that the elements of both the financing scheme and the funding scheme of the 'transport infrastructure delivery and operation/maintenance system' (BENEFIT Framework) are directly associated with risk and risk perceptions. This is further explained in the following sections, where indicators for the financing and funding schemes are presented in more detail.

3.4.1 Financing Scheme Indicator: capturing the characteristics of financing schemes for transport infrastructure projects

Aristeidis Pantelias and Kay Mitusch

Financing transport infrastructure usually involves large sums of money as projects tend to be large in size. Financing sources vary and so do the requirements or expectations of financiers. In this section, an indicator that captures the risk-return profile of transport infrastructure financing schemes is presented, based on a systematic approach that evaluates the contribution of different financing sources. The approach is based on the consideration of both publicly and privately financed projects. A fundamental underlying assumption is that projects delivered with lower risk and lower cost will always be preferable to the alternative.

3.4.1.1 Public versus private project delivery

Financing schemes, and particularly their determinants, will depend on whether a project is developed by a public or private sponsor. The following four cases can be distinguished:

1   project developed by a public agency;
2   project developed by a publicly owned company or SPV (possibly corporatised);

3   project developed as a PPP;
4   project developed by private company without any government support or contract.

In the case of a purely public project, the infrastructure is developed by the relevant level of government (e.g. line Ministry), as is the case for most schools, hospitals, courts of justice, etc. A project in this case is financed by tax money or by public debt issued by that level of government. As government debt or tax collection is rarely ring-fenced for specific uses,[1] it is impossible to determine which part of the financial sources of that level of government is used for the financing of a particular project. Thus, it is hard to determine what the financing scheme of a public project actually is. In this case, the convention is to take the 'financing scheme' of that level of government itself as being representative of the financing scheme of the project. Thus, if the budget of this level of government is 80 per cent tax-financed and 20 per cent debt-financed, it will be assumed that the public project is 80 per cent public equity-financed and 20 per cent public debt-financed, and the government level's debt rate will be applying to both debt and equity.

Note that, whereas the financing scheme of a purely public project is a somewhat artificial concept, the *funding scheme* may be as important as for private companies, even if some part of the funding is also tax-based. To add more clarity to this distinction, *financing* is not the same as *funding*: the former corresponds to raising capital at the beginning of a project to pay for its development costs, such as construction costs; while the latter corresponds to the long term financial streams that will support the repayment of the financing of the project and is closely related to its overall economic and financial viability.

In the extreme opposite case, a project can be developed by a purely private company without any government support or contract (case 4). In this case, the project is financed by a private company independent of any government action or agreement. The company is using the overall leverage of its balance sheet to finance all its projects (including the transport infrastructure project). It will, therefore, be equally difficult to discern the financing scheme that is applicable to the project from the balance sheet of the whole company. Hence, like the first case, the convention is to take the financing scheme of the whole company as being representative of the financing scheme of the project.

Note that purely private investments are quite usual in the transport *service* industry, like private bus or train operators, ship owners or airlines. In the United States, there are even privately owned and financed integrated rail freight companies that finance and operate their own railway infrastructure. However, in Europe, purely private transport infrastructure investments are rare. In Europe, the natural monopoly property of transport infrastructure almost always leads to government intervention.

In the third case, a transport infrastructure project is developed as a PPP, that is by a private company (usually in the form of a SPV) that acts based on a contract with (part or whole of) a government. PPPs are often established as means to invest in projects which would have not been possible with the budget available to the public (European Investment Bank, 2005). Thus, PPPs cannot simply draw on the public budget but must arrange their financing schemes on their own. For this reason, PPPs are the leading case for analysing financing schemes. Financing schemes of PPPs differ as much as legal combinations of public and private shares can be designed in a PPP arrangement.

The second case, in which a transport infrastructure project is developed by a publicly owned but legally independent company, is an intermediate case between purely public and PPP project development. Such companies use tax-funded elements as well as financial instruments, which are also used by private companies. Most of the publicly-owned companies are formally incorporated and must fulfil the same legal requirements as private companies, even though their owner is a public body, and, also follow public rules with respect to procurement and other liabilities. In this case, a transport infrastructure project can be developed either on the company's balance sheet (similarly to a purely private company) or as a version of PPP where the company will act as the contracting authority. The financing scheme of the project in each case will depend on the method of development.

3.4.1.2 Financial viability condition for infrastructure projects

The goals of a company (either public or private) managing/operating a transport infrastructure project should coincide with the goals of any other private-sector company: profit maximisation. For the case of a public company developing and managing such a project, the sole goal may not be simply the one of profit maximisation. However, even in this case the company will always have the goal of avoiding uncontrolled losses in the long run. Consequently, a general financial viability condition for a transport infrastructure project – be it private, PPP or purely public – is that no losses are incurred in the long run.

Generally, profit is defined as revenues minus costs. Let costs here be defined excluding interest expenses, then the profit must cover at least the interest expenses. But equity owners must also be remunerated, else they will withdraw their capital from the firm. Thus, a company is viable only if its profit exceeds the sum of (i) interest expenses and (ii) the opportunity cost of equity. This sum is called the *total capital costs* of the company. Hence, viability requires that profit exceeds total capital costs. Dividing by the amount of invested capital, this viability condition becomes:

$$\frac{\text{Profit}}{\text{Invested capital}} > \frac{\text{Total capital costs}}{\text{Invested capital}}$$

The left-hand side of the condition is defined as the Return on Invested Capital (RoIC):

$$RoIC = \frac{\text{Profit}}{\text{Invested capital}}$$

To be precise, the numerator of this fraction refers to the Net Operating Profit after Tax (NOPAT), which is calculated without subtracting interest expenses, while the denominator refers to the book value of the invested capital (Brealey et al., 2011).

The right-hand side of the viability condition can also be replaced by a prominent expression. In effect, the invested capital (denominator) consists of debt and equity. Moreover, total capital costs (nominator) can be decomposed into total debt cost (= the interest expenses of a typical company of this type) and total equity cost (= the opportunity costs of such company's equity). The following notation is introduced:

K_E as the unit cost of equity;
K_D as the unit cost of debt;
E as the market value of the firm's equity;
D as the market value of the firm's debt.

In the above notation, the market values of debt and equity are often approximated by their respective book values.

The invested capital equals $E + D$ the total cost of debt equals $D \times K_D$, the total cost of equity equals $E \times K_E$ and thus the total cost of capital equals: $D \times K_D + E \times K_E$.

Then, the right-hand side of the viability condition becomes $(D \times K_D + E \times K_E)/(E + D)$ and coincides with the Weighted Average Cost of Capital (WACC), which is defined as:

$$WACC > K_E \times \frac{E}{E+D} + K_D \times \frac{D}{E+D}$$

In this equation, the costs of each category of capital (equity and debt) are proportionately weighted according to their amount in the company's financial structure. Thus, the WACC shows how much interest the company should pay for every money unit it finances.

Putting everything together, the viability condition becomes:

RoIC > WACC

In summary, the viability condition is ultimately the ability of the project to meet its expected financial targets, that is to repay in full and in due time its debt investors and in addition to generate adequate returns for the equity investors.

3.4.1.3 Sources of finance and the risk-return appetite of investors/ financiers

By using the same substitutions as in the case of the formulation of the viability condition of a transport infrastructure project, the general goal of profit maximisation can be reformulated as (Mitusch et al., 2015, pp. 11–19):

max(RoIC − WACC × invested capital)

Thus, to maximise the net return, it is also important to minimise the WACC (all else being equal). In addition, the company would also want to maximise the amount of invested capital. This is completely true for certain investments that can be scaled without affecting RoIC and WACC. However, in most cases, RoIC and WACC will not be independent from the size of the investment. The latter is done by finding the optimal financing scheme, that is combination of debt and equity.

However, the theory of corporate finance asserts that the WACC is fixed, given and independent of the financing scheme (Modigliani Miller Theorem). But this 'irrelevance property' rests on many idealised assumptions, such as, that there are no taxes and no transaction costs (e.g. cost of default); that there is perfect information in a very strong sense implying, in turn, that there is no asymmetric information and, thus, that there are no incentive issues, as all such issues would be resolved in endless 'complete contracts' (Brealey *et al.*, 2011).

In reality, financing schemes will be chosen in such a way as to minimise (expected) taxes, transaction costs and disincentives. For example, a prominent trade-off between equity and debt comes up if one introduces just taxes and some transaction costs of default. According to US legislation, interest on debt is tax-deductible, so that a firm has an incentive to reduce its WACC by raising more debt. The WACC formula is then extended to show this 'tax shield' effect (Brealey *et al.*, 2011). However, using more debt increases the probability of default, which, in turn, relates to additional transaction costs. Hence, a U-shaped relationship has been suggested to exist between the percentage of debt (leverage) in a project's capital structure and the WACC. Then, at a certain level of leverage, the WACC is minimised: this is the optimal debt/equity ratio (Tan, 2007).

In reality, there is not just 'equity and debt' but there are different classes of equity and debt (for example, debt of different priorities at liquidation), as well as intermediate types of finance (so-called mezzanine finance). In addition, there are some financial instruments, which are not regarded as part of the capital structure, but still help to reduce the overall WACC. Furthermore, the WACC formula will be extended such that the cost of each category of finance is proportionately weighted according to its amount in the company's financial structure. Examples are insurances and guarantees that, for a price, isolate the company's capital from specific risks. If available, government guarantees to carry certain risks are always welcome as they usually come at a very low (or no) price for the company. In financial theory, these instruments are also called 'credit enhancing' or 'credit substituting' instruments, to emphasise their role of supporting the company's total capital.

The managers of a project will consider all these financial instruments – equity and debt of different classes, mezzanine and 'credit enhancing' instruments – simultaneously, to find a mix that minimises its overall cost of financing. At the same time, the managers will also consider the various sources of finance on which they can draw capital.

As sources of debt, one should at first differentiate between banks and non-banks, keeping in mind that usually more than 80 per cent of project debt finance comes from banks (Esty *et al.*, 2014). Banks should be further distinguished into leading and non-leading banks. Lead banks take up a rather large portion of a project's debt and, as a pre-condition, require a close and careful look into the project's business plans and other material and may also interfere with the business plans in order to limit overall risks; the management input of a lead bank can be seen as a further means to reduce default risks, and, thus, the WACC. In contrast, a non-leading bank assesses the project more superficially, maybe relying on the leading bank's credit as a positive signal, and takes up only a smaller share of the project's debt. Leading and non-leading banks can also be formally connected in a syndicate. Non-bank credit can be further distinguished as coming from (i) institutional investors (such as pension funds) or (ii) the public (bond issue).

The same distinction applies to equity, where specialist investment funds (e.g. infrastructure funds or private equity funds) would take a role similar to non-bank institutional investors for credit. 'Traditional' equity would be expected from project affiliated sponsors (e.g. construction/operation companies), while other sources can also be encountered (such as equity from banks or from the government).

Sources of specialised risk mitigating instruments are once again the banks and other financial institutions, sometimes even the general public. Government institutions (or government-supported instruments) may also play an important role.

Every element of finance will come at a cost (except perhaps for government support when no financial return is expected). Apart from the risk profile of the project itself, it is the specific risk-return appetite of the financers offering the various financial instruments that determines their cost. Henceforth, 'financier' will be anyone who acts as a source of any type of finance, that is not only of debt, equity and mezzanine, but also of insurances and guarantees.

The risk-return appetite of the financiers can be related to the following parameters:

(i)   financiers' own cost of refinancing with respect to their next best investment alternatives (determining their 'opportunity cost of capital');
(ii) financiers' internal BM (risk diversification motive, specific knowledge about certain financial products or sub-markets);
(iii) financiers' risk perceptions and appetites.

However, the above parameters are difficult to observe in practice due to the commercial and, thus, confidential nature of this information. Consequently, proxy variables have to be used in order to determine the risk appetite of financers and their perception of the riskiness of the project. Such proxy variables are:

• existence of government contribution: if existing, it will directly reduce the part needed from other financers;

- financiers' general operating conditions (i.e. implementation context) such as general market liquidity, market confidence, financial and legal system, political confidence, macro-economic conditions: they will affect financiers' willingness or ability to provide finance;
- type of common financing committed (debt vs. equity vs. mezzanine): committing debt reflects risk aversion, committing equity reflects risk-tolerance, while mezzanine has mixed characteristics;
- project gearing ratio: more equity in the mix denotes higher riskiness and thus the requirement of a bigger equity buffer;
- type of specific risk mitigation instruments (like insurances and guaranties) committed: the use of such instruments is first a clear and precise indicator about the existence and relevance of the corresponding risks (thus informative about the financial profile of the project,). Additionally, the commitment of risk mitigation instruments shows that these risks are considered 'unbearable' (i.e. bearable only at unusually high costs) by usual providers of capital, but as 'normal' by specialised institutions that can insure against them at a low price;
- existence of (formal) credit ratings: when available, they may reduce the cost of debt depending on the rating category that the project falls into. They also provide a market signal in terms of third party due diligence that has gone into a project which may help both equity and debt financiers with their respective decisions;
- financing terms: Higher interest rates and Annual Debt Service Cover Ratios (ADSCR)[2] as well as higher required IRRs show higher perceived project riskiness;
- sources of finance used in project's financing structure: in this case, common market assumptions could be used to benchmark investor risk-return appetite, that is institutional investors being very risk averse, banks not being afraid of construction risk, etc.

3.4.1.4 Financing Scheme Indicator

Based on the consideration of all previous concepts, an indicator is proposed that aims to capture the risk-return profile of a financing scheme that is put forward for a transport infrastructure project. The proposed Financing Scheme Indicator (FSI) has the following formulation:

$$FSI = 1 - WACC_{EXP}^{Adj}$$

where:

$$WACC_{EXP}^{Adj} = \frac{WACC_{EXP}}{z}, \; z \in \mathbb{Z}^+$$

and:

$$WACC_{EXP} = K_E \times \frac{E}{E+D+G} + K_D \times \frac{D}{E+D+G} + K_G \times \frac{G}{E+D+G}$$

The following notation is relevant:

$WACC_{EXP}$ is an expanded version of the Weighted Average Cost of Capital (WACC) of the project under consideration;

$WACC_{EXP}^{Adj}$ is the adjusted version of the $WACC_{EXP}$ based on scaling the theoretical costs of funds assigned to the different sources of equity and debt;

K_E is the unit cost of the equity committed to the project. In the absence of information that would enable the use of the capital asset pricing model (CAPM) for its estimation, theoretical values can be used that are linked to the source of this equity contribution;

K_D is the unit cost of debt committed to the project. In the absence of information about loan interest rates or bond coupons, theoretical values can be used based on the source of this debt contribution;

K_G is the unit cost of public sector funds committed to the project;

E is the market value of the project's equity contribution (in monetary value or percentage);

D is the market value of the project's debt contribution (in monetary value or percentage);

G is the market value of the public sector's funds committed to the project (in monetary value or percentage);

z is the factor that has been used for scaling the cost of funds for the various sources of equity and debt. In general, z is assumed to be a positive integer number ($z \in \mathbb{Z}$).

ADDITIONAL NOTES:

1   The FSI considers the impact of the financing structure to the project based on the cost of funds of the various sources of capital. It is based on a variation of the concept of the Weighted Average Cost of Capital (WACC), as explained previously.

2   All projects, whether purely publicly financed or privately (co)financed (e.g. PPP), are assumed to be drawing their financing from the following three sources of capital: (private) equity (E), (private) debt (D) and public sector funds (G). These sources of capital are divided into additional subcategories.

3   The proposed expanded version of the WACC is calculated by using theoretical values for the cost of funds of the different sources of capital committed to the project due to lack of more detailed information. Upon existence of relevant information, the Financing Scheme Indicator can be undertaken by a proper estimation of the project's WACC. In that case, the

adjusted WACC ($WACC^{Adj}$) would be estimated by dividing the WACC with the highest cost of funds committed to the project.

4   For reasons of simplification (and due to lack of information on loan interest rates and corporate tax rates) the effect of the corporate tax rate[3] on the debt cost of capital is assumed to be already reflected in the value of K_D.

5   For publicly financed projects K_G will be 0, by definition, as for these projects the public sector does not have an expectation of making a financial return and all funds committed to them are usually 'gifted'. Likewise, public sector equity committed to publicly financed projects will also be considered to have a cost of funds equal to 0 ($K_E = K_G = 0$). For privately (co)financed projects, the contribution of the public sector is differentiated depending on whether this is committed as:

- a subsidy, in which case K_G is going to be considered 0, by definition;
- a public sector (true) equity, where the cost of funds should be expected to be high reflecting financial return expectations that would be (on average) commensurate to the expectations of other equity investors in that project; a debt substitution (i.e. government guarantee), when the government is assumed to substitute the original debt facility with contingent government funds using a guarantee. This guarantee is assumed to be provided for a (usually small) premium and consequently the corresponding cost of funds would be:

$$K_G = K_D^{Gov\ Gmt} = 1$$

Overall, the proposed FSI conforms to the fundamental assumption underlying the development of all other indicators in that project structures with lower cost and lower risk should yield indicator values that tend to approach 1. In the case of the FSI, the lower the $WACC^{Adj}_{BEN}$ of a project (i.e. its overall cost of funds), the closer to 1 the indicator value will be.

3.4.1.5 Unit cost of funds for various sources of capital

The various providers of private equity and debt, as well as public sector funds and other types of contributions are assumed to have a varying cost of funds (Mladenovic *et al.*, 2016, pp. 132–133). These are all summarised in Table 3.4.

The values in Table 3.4 can be thought as the expected returns that these investors would anticipate from investing in a transport infrastructure project.

3.4.1.6 Blending of various sources of capital

For projects where more than one source of equity and/or debt have been committed, then the overall cost of equity K_E and cost of debt K_D will need to be estimated by using the following formulas (see Gatti, 2013, for a similar approach):

Table 3.4 Unit cost of funds for various categories of private and public financing

Financing type	Source and description	Unit cost of funds[1]
Equity (E)	Private sector or public sector	K_E
Equity 1	Private equity (PE)-type funds and other short-term financial equity investors	20
Equity 2	Individual affiliated investors (e.g. contractors, operators and other project sponsors)[2]	14
Equity 3	Infrastructure funds and other long-term financial equity investors	11
Equity 4	Commercial banks (equity investment)	8
Equity 5	Public sector (government or similar)	Various[3]
Debt (D)	Private sector	K_D
Debt 1	Debt investors: the general public (tradable bonds), institutional investors,[4] non-leading banks, debt funds	6
Debt 2	Lead banks	5
Debt 3	EIB and other multilateral banks that are mainly self-financing	3
Debt 4	National or international development banks (e.g. EBRD, KfW, etc.)	2
Government support (G)	Public sector	K_G
Gov. sup. 1	Government guaranteed debt	$K_G = K_D^{GovGrnt} = 1$
Gov. sup. 2	Public sector funds/government subsidies	$K_G = 0$

1   The unit cost of funds can be considered as the expected return that these investors would antici-
    pate from investing in the infrastructure project. Theoretical values have been provided in the
    absence of real market data. Values are purely indicative and should not be used for investment
    decisions.
2   Including retail equity investors and/or crowdfunding (if applicable).
3   The following variations are possible:

   • If public sector equity is committed to a publicly financed project, then $K_E = K_G = 0$.
   • If the public sector is participating as a true equity investor in a privately (co)financed project,
     then $K_E^{Public\ sector} = \sum_{i=1}^{N} K_E^i / N$ (i.e. the public sector would expect to earn at least as much
     as the average other investor in this project and would have a commensurate unit cost of
     funds).

4   For example, pension funds, insurance companies, sovereign wealth funds (SWF), etc.

Source: Authors.

$$K_E^{Total} = \sum_{i=1}^{N} K_E^i \times \frac{E^i}{E}$$

and

$$K_D^{Total} = \sum_{i=1}^{M} K_D^j \times \frac{D^j}{D}$$

The following notation is relevant:

K_E^{Total} is the blended unit cost of funds for all equity contributions;
K_E^i is the unit cost of funds for the i source of equity;
E^i is the market value of the equity contribution from source i;
E is the market value of the total equity contribution to the project;
K_D^{Total} is the blended unit cost of funds for all debt contributions;
K_D^i is the unit cost of funds for the j source of debt;
D^j is the market value for the debt contribution from source j;
D is the market value for the total debt contribution to the project.

Figure 3.4 presents in a histogram the FSI (FSI) values of the BENEFIT project case dataset. As noted, FSI addresses PPPs and Public Financed projects uniformly, with:

- FSI → 1 describing projects with a higher contribution of financing from the public sector;
- FSI = 1 representing a project fully financed by the State, and;
- FSI → 0 indicating projects fully supported by private financing.

One would expect PPP projects to have an FSI < 0.5, that is a higher percentage of private funds committed to them. However, few projects classified as PPP demonstrated such a FSI < 0.5. Most cases were geared towards FSI = 1 showing heavy support by the public sector (Roumboutsos et al., 2016d, pp. 31–32).

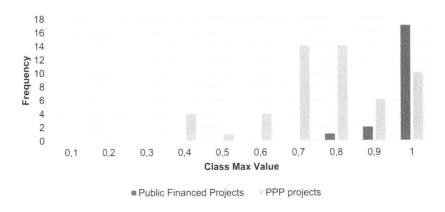

Figure 3.4 Histogram of FSI values of the BENEFIT project case study set
Source: Roumboutsos et al., 2016d, pp. 31–32.

3.4.2 *Funding Scheme Indicator: project revenues and returns*

Joao Bernadino and Athena Roumboutsos

Funding schemes, in the form of both revenue streams and remuneration method, are connected to the feasibility of infrastructure delivery. Even if a project is considered viable from a socio-economic cost-benefit point of view, it is simply impossible to realise unless adequate sources of revenue are identified and mobilised to pay for its costs and support a financing package for its implementation. In other words, while a project may generate enough social benefits to justify its delivery, these may not be adequate to justify its financing as they may not be possible to be captured in monetary terms.

Funding schemes are connected to the potential of securing financing and establishing sufficiently appealing returns (in terms of magnitude and risk) for investors. A fundamental reason a project would not generate sufficient funding streams, even though it may provide more overall value than its delivery costs, is the fact that this value may be dispersed across several actors and effective 'value capture' mechanisms may not be possible. Therefore, funding schemes for infrastructure projects carry significant uncertainty with respect to the ability to capture potential streams of revenue. In effect, funding models 'that rely on annual budget funding cycles to supplement user revenues rarely produce satisfactory outcomes; what counts is secure funding from multiple and diverse sources' (OECD, 2011). To make secure funding from different value recipients possible and available, appropriate institutional conditions and BMs are needed.

The selection of basic factors to characterise funding schemes attempted to be as exhaustive as possible, that is covering the widest possible range of types of funding (either currently in use or with a potential to be used in the future) applied in transport infrastructure provision. An additional criterion was capturing the essential attributes relevant to the effects of funding schemes on PP factors (Vanelslander *et al.*, 2015, pp. 48–56; Roumboutsos *et al.*, 2016b, pp. 100–104; Mladenovic *et al.*, 2016, pp. 126–130)

A funding scheme is characterised by a set of revenue streams originating from or collected for the purpose (function) of a project. It is, therefore, necessary to distinguish the characteristics of each of the revenue streams that comprise it, along with the risk of 'capturing' each one of these. It is, also, necessary to characterise how these revenue streams are distributed among actors. To optimise risk allocation and incentives, funding schemes also comprise remuneration schemes which are not directly similar to the revenue streams. Remuneration schemes are determined with reference to an agent involved in project delivery (e.g. operator) through the re-composition of the project's revenue streams based on given criteria that are appropriate for the optimisation of the contractual arrangement. The characterisation of a funding scheme therefore includes the definition of the remuneration scheme, that is to whom and how the revenues related to the project are attributed.

Finally, the fundamental question related to funding is whether the available funding schemes (revenue streams and remuneration) are sufficient to cover the costs of the project to make it financially viable. This ability for *cost recovery* has significant variation among different modes of transport.

In this context, the funding scheme is considered in the following sections through its dual nature: the revenue streams generated by the project and, separately, as the remuneration scheme that captures revenues and allocates them to the appropriate project stakeholder/agent.

3.4.2.1 Revenue schemes and the Revenue Robustness Indicator

Funding transport infrastructure through equitable charging for the benefits generated by the asset has been the object of discussion in many countries during the last decades. In general, three sources of funding may be considered:

(i) The Government or State through the public budget (collected through taxation, etc.), as project services may benefit society as a whole, thus making the asset a 'public-good'.
(ii) User charges, where the users pay for the use of the infrastructure asset or the consumption of the relevant service (e.g. toll roads, public transit, etc.). In this case, benefits (either in whole or in part) may be attributed to and are paid by the users.
(iii) Charges to (or other input from) indirect 'beneficiaries' (i.e. charges targeted at non-user beneficiaries of the transport investment). Examples include a betterment tax that may be levied on properties that benefit from the construction of community infrastructure.

Revenue schemes concern the way transport infrastructure produces revenue as well as how the respective infrastructure services are charged for. In this context, the issue of charging for and pricing of infrastructure is of relevance. Charging for and pricing of infrastructure has been a topic discussed and studied extensively in national and international policy and research due to its relation to equity and mobility. It is also related to the allocation of demand risk in light of price elasticity demonstrated with respect to different transport services offered.

There are several possible revenue streams originating from the agents identified above. Conceptually, they are applicable to all modes of transport, although they are distinctively applied in different modes and projects. The following revenue streams may be considered:

- public budget generated by general or regional/local taxation, which applies to direct and indirect users of the infrastructure;
- the user-pays approach, which is applied at varying levels across modes based on (i) the ability to cover infrastructure costs; (ii) the transaction costs involved and (iii) public acceptance and equity concerns.

- social marginal cost pricing: the objective is to obtain an efficient use of capacity and to internalise negative externalities in transport activities. While being a policy objective (EC, Transport White Paper of 2001), its applicability has been limited due to practical barriers;
- earmarked funds, which include road use, consumption, local motoring, employer/employee charges, property-related taxes, parking charges/fines, etc. (Ubbels and Nijkamp, 2002). The merits of earmarking in terms of economic efficiency, revenue stability and public acceptability vary widely according to the earmarking schemes applied and their context (OECD, 2011);
- land value capture, which concerns collection of revenues by levying fees on parties, which directly benefit from the infrastructure asset or its services, not as users, but in the form of increasing property value (Martinez and Viegas, 2012);
- bundled services or infrastructure. This concerns revenues from commercial and/or other activities that may be bundled with the core infrastructure asset or service and may exert additional rents by exploiting exclusivity rights.

Notably, each of the above revenue streams carries different risks. In common practice, a combination of them may be employed to finance a transport infrastructure project.

Considering these conditions, the revenue scheme is characterised by the level of robustness of the combination of revenue streams. Hence, the proposed indicator, Revenue Robustness Indicator (RRI), is formulated as follows:

$$RRI = \sum_{i}^{n} (1 - risk_i) * cov_i$$

where i represents the ith revenue stream, with risk factor, $risk_i$, representing a cost coverage, cov_i (per cent).

3.4.2.2 *Remuneration schemes and the Remuneration Attractiveness Indicator*

Remuneration generally refers to payments against previous expenditures. In the context of transport infrastructure projects such expenditures may be related to initial construction costs and operational/maintenance costs. Remuneration may, thus, be considered as payments made to the project stakeholders/actors that have incurred or are currently incurring project-related costs (or seek to recuperate their investment). Consequently, remuneration schemes are always related to the stakeholder/actor to whom such payments are attributed. Finally, remuneration schemes are also related to the method of capturing project revenue streams to restructure them as payments towards the relevant stakeholders/actors. When considering remuneration schemes for transport infrastructure projects, three basic options may be identified:

(i)   Real Tolls, by which users pay the operator directly for the use of the asset/service;
(ii) Shadow Tolls, by which the Government/State compensates the operator per unit of asset utilisation or service usage;
(iii) availability fees, by which a periodic payment is made by the Government/State to the operator, based on service performance criteria, with asset availability usually being one of them.

Different variations and combinations of the above options usually exist in practice. In addition, based on the above definitions, remuneration schemes may be categorised as:

- demand-based, when the remuneration is based on demand for the asset/service (real/direct and shadow tolls/charges); and
- availability- or performance-based, when the remuneration is based on asset availability or service performance criteria, respectively.

Additionally, depending on the origin of the payment, remuneration schemes may be categorised as:

- user-paid (real tolls/direct user charges); or
- government-paid (shadow tolls and availability fees).

While adopting demand/based, user-paid remuneration schemes would reduce transaction costs, availability/performance-based, government-paid schemes are often adopted (or partially adopted) when:

- the project exploitation does not generate (enough) revenues to pay for its costs and additional revenues are needed. Reasons for insufficiency of funds may include standardising toll rates nationally for regional equity or public acceptance (ITF, 2013) or the level of demand and willingness-to-pay being simply insufficient to cover the costs.
- the repayment of project costs through revenues generated by the project would induce sub-optimal risk-distribution and incentives to the project stakeholders.

In effect, when revenue risk is fundamentally related to demand risk and when the project has limited power to mitigate that risk (whereas the public party may have more power to influence demand), it becomes more rational to attribute such risk to the public party (Roumboutsos and Pantelias, 2015).

While the above may be considered payment streams directly related to the project, certain revenue streams, as described previously, may also be directly employed as payments such as those generated by activities bundled in the provision of the infrastructure (e.g. commercial activities, advertising, etc.).

In addition, the remuneration scheme may concern a synthesis of the above and, therefore, may refer to several remuneration methods. Each method is associated with the respective risk and corresponds to a level of cost coverage. Obviously, the less risk associated with the remuneration stream, the more attractive it will be for investors. Hence, the formulated indicator for the remuneration scheme is termed Remuneration Attractiveness Indicator (RAI) and is described by the following relation:

$$RAI = \sum_{i}^{n} (1 - \text{risk}_i) * \text{cov}_i$$

where i is the *ith* payment stream, with risk factor, *risk$_i$*, representing a cost coverage, *cov$_i$* (per cent).

Risks associated with the remuneration scheme are a function of the charging regime and various contextual aspects. For example, Perkins (2013) argued that, in the case of roads, toll regimes based on a simple structure with a flexible review possibility in countries where tolls are well established are less risky and comparable to availability fees. However, risk is also a perception and following the GFC of 2008/9, a clear preference for availability-based, government-paid remuneration schemes has been demonstrated (see Figure 3.5).

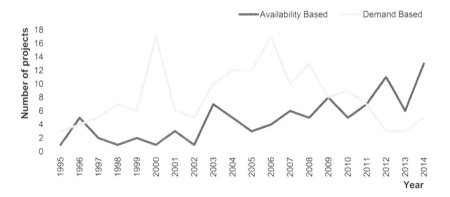

Figure 3.5 Availability-versus demand-based remuneration schemes over time (PPP financial close)

Source: Mladenović et al., 2016, pp. 208–212.

3.5 The importance of transport infrastructure mode

Thierry Vanelslander and Eleni Moschouli

This section seeks to describe the transport mode typology through an indicator influencing PP (outcomes). The framework relies on a reduction methodology that ensures the reduction by using a set of project cases for which parameter values for a long list of possible indicators have been collected. By applying the proposed methodology, the final choice of transport mode indicators suitable for the BENEFIT Framework is made.

3.5.1 From a long list to a short list: the methodological framework

Moschouli and Vanelslander (2016) presented a long list of factors/variables that could be adopted as indicators for various dimensions of the transport mode context typology. From the long list of dimensions and respective indicators proposed, three main dimensions were finally kept as the most critical ones for the funding and financing of transport infrastructure. The selection of these three dimensions is made based on the literature review, authors' experience, data availability and interrelation with other key elements of the BENEFIT Framework that affect the performance of transport infrastructure projects.

- Investment: Total investments/costs is one of the indicators suggested under the first dimension 'investments'. This indicator is also taken into consideration by the financing scheme that examines the sources of financing, the cost of capital and the risk of each of the financing schemes. 'Lifetime' is the second sub-dimension included under 'investment' and is measured through the 'contract duration'. This indicator is not overlapping with any other typology. However, 'contract duration' is an important indicator affecting success of transport infrastructures, in combination with 'availability'.
- Users: Five key performance indicators are suggested to measure the 'performance' of transport infrastructure: 'reliability', 'availability', 'maintainability', 'safety' and 'security'. These indicators can affect the demand of the users for the infrastructure (either passengers/individual users, freight users or both). Risk allocation, assessment and mitigation are also taken into consideration. Regulatory risk is taken into consideration by the implementation context.
- Market strength/competitive position: this includes the 'location', the 'level of integration' and the 'level of exclusivity' of the project.

The full list of factors/variables considered under each dimension is presented in Table 3.5.

A good methodology to deal with the analysis/examination of only a limited number of cases with limited information available is Qualitative Comparative Analysis (hereafter QCA). This method keeps the middle between a qualitative and a quantitative approach. Previous research has shown the appropriateness of qualitative methods, surveys and factor analysis for assessing equifinality[4] (Gresov and Drazin, 1997). Set-theoretic methods such as QCA offer a systematic approach that at the same time can examine extensive numbers of different combinations of indicators/variables, but does not disaggregate the case as a variable-based approach would (Fiss, 2008).

Table 3.5 Transport mode typology dimensions and indicators

Dimensions	Sub-dimensions	Indicators
Investment	Level of sunkness of investment	Non sunk/sunk investments
	Investments/ costs	Construction – CAPEX Maintenance and operation – OPEX
	Lifetime	Project/infrastructure (investment) life cycle Contract duration/infrastructure life
Users	Users	Number of freight vehicle-kms Number of passenger vehicle-kms
	Operational flexibility-continuity	Rerouting
	Performance	Reliability (% time of disruptions) Availability (% of days in year) Maintainability (% of not available) Safety and security (cost of accidents)
	Capacity	Vehicles/hour
	Risks	Demand risk Regulatory risk Financial risk Revenue risk Design risk Construction risk Maintenance risk Exploitation risk Force majeure Climate change risk

Market strength/ competitive position	Location	Type of connection:
		• Interurban
		  – International
		  – National
		  – Regional
		  – Local
		• Urban
		• Node within a node
		• Link within a link
		• Node
		• Link

	Level of regulation-deregulation	Technical harmonisation Noise and pollution emissions • Noise level per mode • % of emissions per mode Pricing: degree of tariff freedom State grants

continued . . .

Table 3.5 Continued

Dimensions	Sub-dimensions	Indicators
		• Grants to cover infrastructure costs • Grants/subsidies to cover the operation of the infrastructure Market Liberalisation Index
	Level of integration	Physical integration Operational integration Governance integration ICT integration
	Level of exclusivity	Natural or induced monopoly and influence of the transport network

Source: Moschouli and Vanelslander (2016).

QCA involves the imposition of theoretical and substantive knowledge in examining imperfect evidence. In this regard, set-theoretic methods are faced with the same issues of causal inference as all other methods that use non-experimental data. QCA is not useful in very small-N situations (e.g. less than 12 cases). Effective use of QCA depends on the ratio of cases to causal conditions. QCA is currently also not designed to perform truly dynamic longitudinal analyses (Fiss, 2008). All variables are transformed into fuzzy sets using the 'direct' and 'indirect' method of calibration, in order to apply fuzzy set QCA (fsQCA) (Ragin, 2000; 2008). The rescaled measures range from 0 to 1 and are tied to their respective membership thresholds and crossover points. In fsQCA, an important aspect relates to modelling the absence of the outcome. In this case, that means modelling the absence of high performance; note that this is different from modelling causes leading to low performance. It should be also pointed out that QCA thus allows for Causal Asymmetry,[5] a concept foreign to correlational methods that always perceive causal relations in symmetric terms (Fiss, 2008). The minimum number of cases needed to do a reliable QCA is approximately 12.

3.5.2 Case selection and data collection

Based on the total BENEFIT project case dataset (see Appendix), which contains 56 PPP and 31 purely public cases, the number of variables (indicators) from Table 3.5 was reduced. Only the input variables for which a sufficient number of cases have values available, were retained.

Based on the above-mentioned variable (indicator) selection process, the final analysis sample contained 34 PPP projects and 19 public projects. Case data were collected through desk research and interviews and streamed lined between cases on a qualitative scale, which, for the purpose of the fsQCA, was expressed on a numerical scale of 0 to 1.

3.5.3 Operationalisation of the Transport Mode Indicator

FsQCA was applied to three dataset partitions: (1) only on PPP projects, (2) only on public projects and (3) on PPP and public projects together. The variables that were obtained in all three cases were 'availability' and 'reliability'. These two variables are included in one of the main typology dimensions, that is 'users'. There is one alternative result from the analysis of PPP projects, which combines as key variables 'duration' and 'availability'. These two indicators span two of the three main dimensions, that is 'investments' and 'users', respectively.

Regarding the resulting combination of 'availability-reliability', these two indicators are fairly important for the transport mode element. This is particularly true for 'availability', because it has a direct impact on the funding and financing of transport infrastructure. Availability measures the time transport infrastructure is available to the users over a unit period. Performance indicators, such as availability, are stated explicitly in many project contracts, and may affect the risk perception of investors because of its implications to the revenue-generating capacity of the project and thus its ability to deliver forecast returns.

Reliability is the other indicator coming back as an outcome from the fsQCA analysis. Reliability reflects the degree of trust for each mode's ability to deliver its intended service and is measured as the percentage of time of disruptions during operation. Reliability is a key factor in users' mode choice.

The alternative combination of results, which was attained from the analysis of solely PPP cases, combines as key variables 'contract duration' and 'availability'. Notably, the 'availability' indicator is prominent again, as one of the two indicators of the combination. The 'contract duration' indicator is used to measure the dimension 'investments'. In particular, 'contract duration' is the numerator in the ratio 'contract duration/infrastructure life'. Contract duration is also an important indicator of the transport mode context element because it has a direct impact on ownership and the period over which investment returns are anticipated. However, it is also related to potential unforeseen risks due to the incomplete nature

Table 3.6 Case solution paths

	Strongest combination/path of conditions – Intermediate solution	Consistency cut-off	Solution coverage[6]	Solution consistency[7]
Private cases	Reliability*Availability	0.829	**0.919**	**0.829**
Public cases	Reliability*Availability	0.889	**0.941**	**0.889**
Combined cases	Reliability*Availability	0.847	**0.926**	**0.847**

Solution coverage = raw and unique coverage
Path's consistency = solution consistency in this analysis
Source: Authors.

of the respective long-term contracts as well as economic cycles (European Bank of Reconstruction and Development, 2015).

As the combination of 'reliability' and 'availability' concluded on a greater coverage for the combined sample of PPP and public projects, these were adopted to construct the transport mode element indicator. Notably, both 'reliability' and 'availability' were included in the 'performance dimension'.

The qualitative assessment of the indicators is converted to a score in the range [0,1] and combined to describe the overall indicator for reliability/availability (IRA):

$$IRA = (1 + I_R) * \frac{(1 + I_A)}{4}$$

where:

i_R is the Reliability Indicator
I_A is the Availability Indicator
IRA   overall Reliability/Availability Indicator

What makes IRA as a resulting composite indicator different from other indicators is the fact that its sub-indicators can take discrete (e.g. 0, 0.5, 1) and not continuous values from 0 to 1. Thus, when assessing the reliability and availability of a transport infrastructure, it can be considered that:

(i)   reliability was improved fully in line with expectations or even more (value 1);
(ii)   reliability was improved partially in line with expectations (value 0.5); and
(iii)   reliability was not improved or only marginally improved (value 0).

A similar approach is expected for 'availability'. Notably, this was a required concession as quantitative data for the reliability indicator were not always available. Stakeholders could only assess if reliability (availability) was better, in line or less than expectations.

3.6 What do system indicators say?

Athena Roumboutsos

3.6.1 A multi-method analysis approach

By developing and constructing indicators to represent the key elements of the BENEFIT Framework, project case information is codified and transferred into a numerical space. This allows for the application of additional tools of analysis and, also, the investigation of an intuitive assumption: that there is never a single factor that leads to a specific result, but a combination of factors that function and interact with each other.

Notably, all analysis tools and methodologies present their own limitations with respect to: the number of observations needed, the number of variables that can be handled, the level of accuracy required, the variation in values that is allowed for the variables considered and many other factors that may influence the ability to interpret their results and make robust and valid inferences and generalisations based on them. To this end, the application of a multi-method analysis approach allows for richer and more reliable conclusions to be drawn by combining their respective findings and thus alleviating some of their short-comings.

Three such analysis methods are selected for their varying level of reliance on the numerical values of the indicators: fuzzy set Qualitative Comparative Analysis (fsQCA, Roumboutsos *et al.*, 2016b, pp. 152–200; Mladenovic *et al.*, 2016, pp. 152–186); Importance analysis (or Sensitivity Analysis, Roumboutsos *et al.*, 2016b, pp. 201–211; Mladenovic *et al.*, 2016, pp. 187–197) and econometric analysis (Roumboutsos *et al.*, 2016b, pp. 212–231; Mladenovic *et al.*, 2016, pp. 198–205).

All three, start from a different theoretical basis and, hence, approach the research question from a different perspective. Moreover, the conceptualisation of causality in each one of them is different. Econometric analysis searches for causality based on correlations and studies the independent effect of some variables on the dependent variable, while controlling for other variables. FsQCA looks for the combination of conditions that is necessary or sufficient to explain a certain outcome (INUS[8] as concept of causality) and hence focuses strongly on how different conditions in combination interact to bring about a certain outcome. Importance (or Sensitivity) Analysis takes the middle ground. It is conceptually based on Bayesian Networks and maps out cause-and-effect relationships among key variables under conditions, while estimating the marginal log-likelihood of proposed relationships.

By employing these three methods the full set of variables (indicators) may be exploited. The fsQCA does not include the Reliability/Availability (IRA) and the Revenue Support (RSI) indicators, the latter being of high importance as it includes the 'Level of Coopetition' or the control over traffic the project may have, as these indicators did not present sufficient variance for the method. The econometric analysis does not include the InI as it was found to be highly correlated to the Financial-Economic (FSI) indicator. Finally, the Importance analysis includes only the factor 'Level of Coopetition' from the constituents of the Revenue Support (RSI) indicator. In addition, the methods use different parts of the dataset. FsQCA used the inauguration project snapshot for the assessment of Cost and Time performance. A subsequent snapshot (when available) was used for the assessment of traffic and revenue performance. Importance analysis and econometric analysis used all the snapshots excluding the award snapshot that represents the planning stage. Notably, Importance analysis and econometric analysis consider all snapshots as observations of equal value to the analysis, while fsQCA being closer to the actual cases, cannot use the same case twice (or more times). FsQCA provides findings based on pre-selected 'paths' that include combinations of indicators. Finally, all methods do not treat all outcomes independently. The econometric analysis identifies a small correlation between cost and time to completion and, therefore, also considers these outcomes in combination.

As in every analysis, results are influenced by the initial assumptions made. For example, no causality may be found with respect to actual vs forecast traffic if the traffic forecast was overestimated as an input to begin with. This consideration influences mostly semi-qualitative methods (e.g. fsQCA).

3.6.2 Combining findings

Keeping in mind the theoretical background of each analysis method and its respective limitations, the combined interpretation of findings provides richer and more comprehensive conclusions. These are discussed herewith and their summary is presented in Table 3.7. FsQCA assumes 'paths' (combinations of indicators) and validates them for the number of cases in the sample that may be explained by a specific 'path'. The 'path' explaining the greater number of cases is presented in Table 3.7. However, all results are considered in the analysis (Roumboutsos *et al.*, 2016c, pp. 31–38).

The analyses were conducted using different partitions of the sample: the sample as a whole, the sub-sample of PPP projects and the sub-sample taking into account the GFC. Notably, when considering the last sample, the continued impact of the GFC (see Figure 3.2) should be taken into consideration. As expected, results are not available for all samples by all analysis methods.

3.6.2.1 Cost to completion

The econometric analysis identified that the GFC had a definite negative impact on the potential of projects to reach cost to completion targets. A marginal increase in the FEI significantly improves the probability of achievement of these targets (approx. 55 per cent depending on the econometric model applied). This is reinforced by the fact that the fsQCA produced only one (1) path which included a low value of the FEI and applied to only five (5) cases successfully with respect to the cost to completion outcome. The other paths with positive results included a high value of the FEI whether before or after the 2008 year-mark.

The other indicator identified as important is the InI. Notably, this indicator was not included in the econometric analysis, due to issues mentioned previously. However, a high value of InI is present in all combinations of the fsQCA and ranks high in the results of the Importance Analysis. The InI is also found in the path combination including the low value of the FEI.

In addition, the fsQCA provides repetitively one additional indicator with a positive impact: a high value of the FSI, referring to projects which are either delivered by the public sector or that have been heavily supported by it. Remarkably, a high FSI was present in results from the PPP sample. Here, also, the combination of high values of the CSI and InI are supporting the ability to achieve cost to completion targets. In the econometric analysis, the FSI was not found to be significant. However, the indicator carried a 'positive sign' indicating a trend towards higher values in order to achieve cost to completion targets. The Importance analysis found the FSI to be independent of the results in all sample cases.

In conclusion, the comparative analysis highlights the combined importance of high values of the GI, CSI and RSI and the significant impact of the FEI, which may be counter-balanced by high values of the Institutional (InI) and Financing Scheme (FSI) Indicators.

3.6.2.2 Time to completion

The econometric analysis identified that higher values of the GI and lower values of the RAI improve the possibility of delivering projects 'on-time'. As in the case of 'cost to completion', lower values of the FEI have a negative impact. Notably, this effect was found to be comparatively less significant (24.4 per cent). A counter-balancing effect is provided by the GI; a marginal increase in the GI increases the probability of being 'on-time' by 18.3 per cent.

Further to the above, all analyses found that a low value of RAI increases the probability of achieving time targets. Notably, a low value of the RAI corresponds to high risk (demand-based) remuneration schemes, which provide an incentive to start project operation 'on-time'. The Importance analysis shows the incentive (RAI) to be even more pronounced after the crisis for the PPP sample in particular. The fsQCA has found that a low value of RAI in combination with positive implementation conditions (FEI), together with high values of GI and CSI lead to achieving time targets with a high value of the FSI.

Confirming the above, the Importance analysis found GI, RAI and InI to be influencing the potential to achieve 'time to completion targets'. The factor LoC was also identified as important in all cases.

There is agreement in all three analyses that the GI is important in achieving 'time to completion'. A low Remuneration Attractiveness Indicator (RAI) will drive a project to be completed on time according to the econometric analysis, while structures featuring a high Level of Coopetition (LoC; such as bridges and tunnels) will most likely be completed on time according to the Importance analysis. Low GI and low CSI values will impact negatively the potential to achieve time targets according to the fsQCA.

3.6.2.3 Actual versus forecast traffic

The analysis with respect to actual vs forecast traffic is heavily influenced by the initial system assumption that input is correct (i.e. that traffic forecast has been correctly estimated) and that non-achievement of initial estimates is due to the negative impact of a combination of indicators. This assumption has been challenged in the specific project dataset (as in most similar datasets, see, for example, Flyvbjerg *et al.*, 2004). This is evidenced by the fact that the Importance analysis was not able to provide significant results for traffic goals, while the 'paths' identified by the fsQCA are only valid for a limited number of projects.

Considering available results, both the fsQCA and econometric analyses agree that a high Remuneration Attractiveness Indicator (RAI) value and a high CSI value are important in achieving traffic goals. A positive Implementation context (high values of InI and FEI) is also important. Moreover, the econometric analysis highlights the importance of the FEI in achieving traffic targets, as a marginal increase in FEI improves the probability of achieving traffic goals by 72.6 per cent, while a similar increase in the RAI only improves results by 14 per cent.

Furthermore, the significance of a low Remuneration Attractiveness (RAI) and Cost Saving (CSI) Indicators values are reinforced by the fsQCA as factors for not achieving traffic goals when in poor implementation context. A high value of the GI and InI, as per the fsQCA, is found in combinations of the above indicators.

With respect to the impact of the GFC, the value of the Remuneration Attractiveness Indicator (RAI) seems to be important in achieving traffic goals, as well as the CSI (econometric analysis). While the contribution of a high RAI (e.g. availability fees) may be almost self-evident, the importance of the CSI requires further consideration. More specifically, it should be noted that the CSI does not only include factors connected to construction but also factors connected to the capability to operate and the 'maturity' of the contracting authority. In addition, factors related to construction cease to be relevant following project inauguration. A low Remuneration Attractiveness Indicator (RAI) value combined with low GI and a high value of the FSI are reasons for not achieving traffic goals according to fsQCA.

3.6.2.4 *Actual versus forecast revenues*

The Remuneration Attractiveness Indicator (RAI) is key in achieving revenue targets. Econometric and fsQCA analyses also highlight the influence of the RRI. However, this indicator includes cost coverage and, therefore, reflects on its own the level of achieving revenues (or not).

Other indicators found in the positive combinations of fsQCA are GI, CSI and the InI. Also, the FSI seems to provide a positive contribution. Finally, the importance analysis identified the importance of the LoC (factor of the RSI) as significant with respect to the impact of the GFC on achieving revenue targets.

3.6.2.5 *Observations and remarks*

From the preceding discussion, there is *no single indicator* that is dominant in driving the attainment of different outcomes but rather combinations of indicators. Additionally, there is *no single combination of indicators* that can secure the successful attainment of all project outcome targets simultaneously. Finally, the implementation context indicators (FEI and InI), which are external to the project, are always significant.

Table 3.7 Summary of findings

Sample	Method	FEI	INI	GI	CSI	RSI	FSI	RRI	RAI	IRA
Cost to construction completion										
Entire sample	fsQCA									
	Importance analysis		***	**	*	LoC ****				
	Econometrics analysis	****	*	**	***					
PPP sample	fsQCA			+	+	+		+		
	Importance analysis		****	**	*	LoC***				
Crisis sample	fsQCA	Before		+	+			–/+		
		After	+	+	+			+		
	Importance analysis	Before		****	***	*	**			
		After			****	**	*	***		
	Econometrics analysis	****	*	**	***					
Time to construction completion										
Entire sample	fsQCA									
	Importance analysis		***	***		LoC*			**	
	Econometrics analysis	****	+					** (–)	*	
PPP sample	fsQCA					+				
	Importance analysis		***	****		LoC*			**	
Crisis sample	fsQCA	Before								
		After			+			+		
	Importance analysis	Before		****	***		LoC*			***
		After			**	**				*	
	Econometrics analysis	****	***							

continued . . .

Table 3.7 Continued

Sample	Method	FEI	INI	GI	CSI	RSI	FSI	RRI	RAI	IRA
Actual vs forecast traffic										
Entire sample	fsQCA	****		**	+				+	
	Econometrics analysis							***	*	
PPP sample	fsQCA		+	+	+		+		+	
Crisis sample	fsQCA	Before		+	+					
	Econometrics analysis	****		**				*** (–)	*	+
Actual vs forecast revenues										
Entire sample	fsQCA		+	+			–	+		
	Importance analysis							****	****	
	Econometrics analysis							****	***	**
PPP sample	fsQCA		+					+		
	Importance analysis								****	
Crisis sample	fsQCA	Before After		+	+			+	+	
	Importance analysis					LoC****			***	
	Econometrics analysis							****	***	**

Legend: '+' and '–' depict high or low values of an indicator for fsQCA result; (–) depict positive influence as the value of the indicator decreases for econometrics analysis; the number of asterisks (*) depict the relative importance of each indicator as identified in importance and econometrics analysis.

Source: Authors.

The last remark highlights the difficulty in ultimately achieving all project outcomes. A typical example is the Remuneration Attractiveness Indicator (RAI): a low value will drive successful 'on-time' delivery, while a high value will tend to support traffic outcomes. Demand-based remuneration schemes (low RAI) do not support the potential of reaching forecast traffic, especially in an adverse implementation context (low FEI and InI values). However, the presence of the other indicators in achieving revenue targets should not be neglected, as in combination they may be able to 'balance' lower values of RAI. Furthermore, the strategy of selecting a remuneration scheme with a high RAI value to improve asset utilisation (i.e. traffic) can only be achieved under positive government budgetary conditions. This is also evident by the wide variety of countries seen to be adopting a high RAI value policy, that is availability-based remuneration schemes after the GFC (Roumboutsos *et al.*, forthcoming).

Considering the contribution of each indicator, the following remarks are put forward:

- Governance (GI) has a positive influence on outcomes and justifies the efforts placed in conducting a transparent and competitive procurement process but, also, highlights the significance of contractual flexibility.
- A high value of the InI has a positive contribution to all outcomes, suggesting that governments need to strengthen their institutions, especially considering their limited control over the financial-economic context. This finding also justifies efforts placed in strengthening EU institutions.
- The CSI contributes positively to all outcomes apart from 'time to completion', where trade-offs between time and cost to completion are pronounced.
- The Remuneration Attractiveness Indicator (RAI) is behind time and cost trade-offs. Demand-based and user-paid remuneration schemes drive time to completion, while availability-based and government-paid remuneration schemes provide more secure traffic and revenues to the project.
- The RRI, as expected, is characteristic of expectations with respect to meeting revenue targets, as it expresses a project's cost coverage and expected risk. More specifically, high values of RRI suggest greater potential of reaching revenue targets.

It is also interesting to note that the achievement of the revenue target is not dependent on the FEI. However, revenues in all cases are dependent on traffic and traffic has been identified to be influenced by the FEI. Furthermore, governments tend to secure revenues required for the operation of transport infrastructure or adjust maintenance schedules to fit available budgets. Recent reports on the level of infrastructure maintenance tend to support this assumption.

Finally, it is worthwhile commenting on the contribution of the FSI. High values of the FSI correspond to projects financed by the public sector or private projects with a high contribution (or support) from the public sector. All three analysis streams showed that positive outcomes were always related to high values of the FSI. However, the econometric analysis showed a positive probability of meeting time to completion and revenue targets connected to a low FSI, albeit with low significance.

3.7 Conclusions

Expressing the 'transport Infrastructure delivery and operation/maintenance system' (BENEFIT Framework) through its key elements in quantifiable indicators allows for further considerations. The chapter presented the theoretical underpinnings behind identifying key factors that may be included in each system element and its corresponding indicator. This chapter also presented detailed information on indicator formulation and the validation process that followed. The latter was limited, in all cases, by the availability of data. Notably, data availability is a typical research constraint (Chen *et al.*, 2016). However, this constraint led to the construction of indicators that could be used by all stakeholders (with varying access to project information) and during all project phases, including phases when all project information has not been concluded yet, such as during the planning phase.

The ability to transfer project information into an 'indicator space' extends the potential of explanatory analysis in this field of practice. It allows the consideration of multiple factors which, in combination and based on their respective values, become facilitators or barriers to the achievement of project goals. The observations stemming from the comparative analysis of the results of the multi-method analysis approach followed has justified this effort.

All three streams of analysis are based on the BENEFIT project case dataset. A different number of project cases exists per mode of transport, while the impact of the GFC on different modes has been disproportionate. Under the above considerations, the findings from the various indicator analyses need to be carefully scrutinised in light of the results from the qualitative analysis per mode that was presented in Chapter 2. The importance of indicators in the achievement of the respective outcomes is then assessed for each mode. Differentiations in indicator analyses findings are then meaningful as these describe the differentiation between modes, as is discussed in the next Chapter 4.

3.8 Notes

1   Exceptions exist, such as the Highway Trust Fund in the United States, which is funded by gas tax and whose funds are earmarked for transport project development (https://www.fhwa.dot.gov/highwaytrustfund/).

2   The ADSCR is a measure of a company's sustainable liquidity level. A project manager should always have a look at the Annual Debt Service Coverage Ratio:

$$ADSCR = \frac{\text{Annual Net Operating Cash Flow}}{\text{Total Debt Service}}$$

This is a measure of cushion between debt service and cash flow available for debt service per year. A too high ratio of debt may lead to default when the project's cash flow is not sufficient to service debt. This would negatively affect creditworthiness. Thus, a management rule is that ADSCR should never fall below a certain level deemed acceptable (min $ADSCR \geq 1x$.)

3   Usually $K_D = i \times (1 - T)$ where i is the interest rate and T is the corporate tax rate applicable (Brealey *et al.*, 2011).

4   Equifinality: An outcome can be elucidated by multiple, mutually non-exclusive (paths of) conditions.

5   Causal asymmetry: The presence of the outcome may have different explanations than its absence (quoted from Verhoest *et al.*, 2014).

6   'Coverage' is the relative importance of different paths to an outcome (Fiss, 2008).

7   'Consistency' is the proportion of observed cases that are consistent with the pattern (Fiss, 2008).

8   Insufficient but necessary parts of a condition which is itself unnecessary but sufficient.

3.9 References

Abednego, M.P., and Ogunlana, S.O. (2006). Good Project Governance for Proper Risk Allocation in Public Private Partnerships in Indonesia. *International Journal of Project Management*, 24(7), pp. 622–634.

Amit, R., and Zott, C. (2001). Value Creation in E-Business. *Strategic Management Journal*, 22, pp. 493–520.

Anderson, R. D., Mackoy, R. D., Thompson, V. B., and Harrell, G. (2004). A Bayesian Network Estimation of the Service–Profit Chain for Transport Service Satisfaction. *Decision Sciences*, 35(4), pp. 665–689.

Azhar, N., Farooqui, R. U., and Ahmed, S. M. (2008). Cost Overrun Factors in Construction Industry in Pakistan. First International Conference on Construction in Developing Countries (ICCIDC–I), *Advancing and Integrating Construction Education, Research & Practice*. 4–5 August 2008, Karachi, Pakistan.

Bain, R. (2009). Error and Optimism Bias in Toll Road Traffic Forecasts. *Transportation*, 36(5), pp. 469–482. DOI:10.1007/s11116–009–9199–7

Bain, R., and Wilkins, M. (2002). *Traffic Risk in Start-Up Toll Facilities. Infrastructure Finance*. Standard and Poor's.

Borgonovo, E. and Plischke, E. (2016). Sensitivity Analysis: A Review of Recent Advances. *European Journal of Operational Research*, 248(3), pp. 869–887.

Bowman, C., and Ambrosini, V. (2000). Value Creation vs. Value Capture: Towards a Coherent Definition of Value in Strategy. *British Journal of Management*, 11(1), pp. 1–15.

Brealey, R, Myers, S., and Allen, F. (2011). *Principles of Corporate Finance* (10th edition), Berkshire, UK: McGraw Hill.

Burkhart, T., Krumeich. J., Werth. D., and Loos, P. (2011). Analyzing the BM Concept – A Comprehensive Classification of Literature. *Proceedings International Conference on Information Systems*, 32, pp. 1–19.

Calvo, M., and Cariola, E. (2004). Analysis of the Privatization Process of the Water and Sanitation Sector in Chile.Draft Working Document prepared for the United Nations Research Institute for Social Development Project on Commercialisation, Privatisation and Universal Access to Water, September 2004, Geneva.

Chan, A. P., Lam, P. T., Chan, D.W., Cheung, E., and Ke, Y. (2010). Critical Success Factors for PPP in Infrastructure Development: Chinese Perspective. *Journal of Construction Engineering and Management*, 5, pp. 484–494.

Chan, A.P., Scott, D., and Chan, A. P. (2004). Factors Affecting the Success of a Construction Project. *Journal of Construction Engineering and Management*, 130(2).

Chan, D. W., and Kumaraswamy, M. M. (1997). A Comparative Study of Causes of Time Overruns in Hong Kong Construction Projects. *International Journal of Project Management*, 15(1), pp. 55–63.

Chan, S., and Park, M. (2005). Project Coast Estimation Using Principal Component Regression. *Construction Management and Economics*, 23, pp. 295–304.

Chartered Institution of Highways and Transportation Report. (2012). *Infrastructure Funding and Delivery: An Action Plan for Change*. http://www.ciht.org.uk/en/document-summary/index.cfm/docid/54B27334-71E4-4CE8-8EC99D75B61ACE43

Chen, L., and Manley, K. (2014). Validation of an Instrument to Measure Governance and Performance on Collaborative Infrastructure Projects. *Journal of Construction Engineering and Management*, 140(5).

Chen, Z., Daito, N., and Gifford, J. L. (2016). Data Review of Transportation Infrastructure Public–Private Partnership: A Meta-Analysis. *Transport Reviews*, 36 (2), pp. 228–250.

Chhotray, V., and Stoker, G. (2010). Governance: From Theory to Practice. In D. Levi-Faur (ed.), *Oxford Handbook of Governance*. Palgrave Macmillan, pp. 214–247. DOI: 10.1057/9780230583344.

Chui, M., Domanski, D., Kugler, P., and Shek, J. (2010). The Collapse of International Bank Finance During the Crisis: Evidence from Syndicated Loan Markets. *BIS Quarterly Review*, September, pp. 39–49.

Clarke, T. (2004). *Theories of Corporate Governance: The Philosophical Foundations of Corporate Governance*. London: Routledge.

Clarke, T. (2007). *International Corporate Governance: A Comparative Approach*. London: Routledge.

Conteh, C. (2013). Strategic Inter-Organizational Cooperation in Complex Environments. *Public Management Review*, 15, pp. 501–521.

Cooke-Davies, T. (2004). Project Management Maturity Models. In P. W. G. Morris, and J. K. Pinto, eds. *The Wiley Guide to Managing Projects* (Chapter 49). Hoboken, NJ: John Wiley & Sons, Inc. DOI: 10.1002/9780470172391.

Crawford, L.H., and Cook-Davies, T.J. (2005). Project Governance: The Pivotal Role of the Executive Sponsor. *Proceedings of PMI North America*. Toronto: Project Management Institute.

de Bruijn, H., and ten Heuvelhof, E. (2008). *Management in Networks: On Multi-Actor Decision Making*. London: Routledge.

Delhi, V.S.K., and Mahalingam, A. (2013). A Framework for Post Award Project Governance of Public Private Partnerships in Infrastructure Projects. Working Paper Proceedings – EPOC 2013 Conference.

Delmon, J. (2009). Private Sector Investment in Infrastructure: Project Finance, PPP Projects and Risk (2nd edition). Kluwer Law International, The Netherlands.

Demil, B. and Lecocoq, X. (2010). Business Model Evolution. *Search of Dynamic Consistency, Long Range Planning*, 43, pp. 227–246.

Doloi, H. (2013). Cost Overruns and Failure in Project Management: Understanding the Roles of Key Stakeholders in Construction Projects. *Journal of Construction Engineering Management*, 10, pp. 267–279.

Dubosson-Torbay, M., Osterwalder, A., and Pigneur, Y. (2002). E-business Model Design, Classification, and Measurements. *Thunderbird International Business Review*, 44(1), pp. 5–23.

Dyer, J. H., and Singh, H. (1998). The Relational View: Cooperative Strategy and Sources of Interorganizational Competitive Advantage. *Academic Management Review*, 23(4), pp. 660–679.

Eriksson, P. E., and Westerberg, M. (2011). Effects of Cooperative Procurement Procedures on Construction Project Performance: A Conceptual Framework. *International Journal of Project Management*, 29(2), pp. 197–208.

Estache, A., and Strong, J. (2000). The Rise, the Fall, and . . . the Emerging Recovery of Project Finance in Transport. Washington. DC: The World Bank Institute.

Esty, B.C., Chavich, C., and Sesia, A. (2014). An Overview of Project Finance and Infrastructure Finance – 2014 Update. *Harvard Business School Background Note 214–083*, June 2014.

European Bank of Reconstruction and Development. (2015). Doing More, Doing Better: What Would It Take to Double the Right Private Infrastructure Investment in Emerging Markets? Paper Presented at the *PPP, Days*, London.

European Commission. (2011) WHITE PAPER: Roadmap to a Single European Transport Area – Towards a Competitive and Resource Efficient Transport System, COM(2011) 144 final, Brussels.

European Investment Bank. (2005). Evaluation of PPP. Projects Financed by the EIB, Luxembourg. http://www.eib.org/attachments/ev/ev_ppp_en.pdf

European Investment Bank. (2012). An Outline Guide to Project Bond Credit Enhancement and the Project Bond Initiative, Luxembourg. http://www.eib.org/attachments/documents/project_bonds_guide_en.pdf

Evenhuis and Vickerman. (2010). Transport Pricing and Public-Private Partnerships in Theory: Issues and Suggestions. *Research in Transportation Economics*, 30, pp. 6–14

Fiss, P. (2008). Using Qualitative Comparative Analysis (QCA) and Fuzzy Sets. http://professor-murmann.info/index.php?ACT=24&fid=11&aid=56_oPhPCLjduXz7Kt WCeJbr&board_id=1 (Accessed 9 November 2013).

Fitch Ratings (2012). Rating Criteria for Infrastructure and Project Finance, Global Infrastructure and Project Finance, NY, USA.

Flyvbjerg, B., Bruzelius, N., and Rothengatter, W. (2003). *Megaprojects and Risk: An Anatomy of Ambition*. Cambridge: Cambridge University Press.

Flyvbjerg, B., Skamris Holm, M.K., and Buhl, S. L. (2004). What Causes Cost Overrun in Transport Infrastructure Projects? *Transport Reviews*, 24(1), pp. 3–18.

Fouracre, P. R., Allport, R. J., and Thomson, J.M. (1990). *The Performance and Impact of Rail Mass Transit in Developing Countries*. Crowthorne, Berkshire: Transport and Road Research Laboratory.

Galilea, P., and Medda, F. (2010). Does the Political and Economic Context Influence the Success of a Transport Projects? An Analysis of Transport Public Private Partnerships. *Research in Transportation Economic*, 30, pp. 102–109.

Garland R. (2009). *Project Governance: A Practical Guide to Effective Project Decision Making*. London and Philadelphia: Kogan Page Limited.

Garrido Martinez, L. M., and Viegas, J. M. (2012). The Value Capture Potential of the Lisbon Subway. *The Journal of Transport and Land Use*, 5(1), pp. 65–82.

Gatti, S. (2013). Project Finance in Theory and Practice: Designing, Structuring and Financing Private and Public Projects. San Diego, CA: Academic Press.

George, G., and Bock, A. (2010). The Business Model in Practice and Its Implications for Entrepreneurship Research. *Entrepreneurship Theory and Practice*, 35, pp. 83–111.

Gresov, C., and Drazin, R. (1997). Equifinality: Functional Equivalence in Organization Design. *Academy of Management Review*, 22 (2), pp. 403–428.

Guo, F., Chang-Richards, Y., Wilkinson, S., and Li, T.C. (2013). Effects of Project Governance Structures on the Management of Risks in Major Infrastructure Projects: A Comparative Analysis. *International Journal of Project Management*, 35, pp. 815– 826.

Hammami, M., Ruhashyankiko, J.F., and Yehoue, E.B. (2006). Determinants of Public Private Partnerships in Infrastructure. *IMF Working Paper No. 06/99.*

Hawkins, G. D. (1981). An Analysis of Revolving Credit Agreements. *Journal of Financial Economics*, 10, pp. 59–81.

Heide, J.B. (1994). Interorganizational Governance in Marketing Channels. *Journal of Marketing*, 58(1).

Hjelmbrekke, H., Klakegg O. J., and Lohne, J. (2017). Governing Value Creation in Construction Project: A New Model. *International Journal of Managing Projects in Business*, 10(1), pp. 60–83.

International Transport Forum (2013). *2013 Annual Summit Highlights – Funding Transport: Session Summaries.* International Transport Forum, Paris.

Izaguirre, A. K., and Kulkarni, S. P. (2011). *Identifying Main Sources of Funding for Infrastructure Projects with Private Participation in Developing Countries: A Pilot Study.* Washington: World Bank.

Kaliba, C., Muya, M., and Mumba, K. (2008). Cost Escalation and Schedule Delays in Road Construction Projects in Zambia. *International Journal of Project Management*, 27, pp. 522–531

Ke, Y., Wang, S., and Chan, A. P. (2010). Risk Allocation In Public-Private Partnership Infrastructure Projects: Comparative Study. *Journal of Infrastructure Systems*, 16(4), pp. 343–351.

Klakegg, O. J., Williams, T., Magnussen, O. M., and Glasspool, H. (2008). Governance Frameworks for Public Project Development and Estimation. *Project Management Journal*, 39(S1), pp. S27–S42.

Knights, D., Noble, F., Vurdubakis, T., and Willmott, H. (2001). Chasing Shadows: Control, Virtuality and the Production of Trust. *Organization Studies*, 22(2), pp. 311–336.

Lee, M. S., and Moore, A. W. (2014). Efficient Algorithms for Minimizing Cross-Validation Error. Machine Learning Proceedings 1994: Proceedings of the Eighth International Conference, Morgan Kaufmann, 190 p.

Li, H., Arditi, D., and Wang, Z. (2012). Factors That Affect Transaction Costs in Construction Projects. *Journal of Construction Engineering and Management*, 139(1), pp. 60–68.

Luo, Y. (2002). Contract, Cooperation, and Performance in International Joint Ventures. *Strategic Management Journal*, 23(10), pp. 903–919.

Mahalingam, A. and Kapur, V. (2009). Institutional Capacity and Governance for PPP projects in India. Lead 2009 Conference, South Lake Tahoe, CA, USA.

Mallin, C.A. (2004). *Corporate Governance.* Oxford: Oxford University Press.

Mallin, C.A. (Ed.). (2006). *Handbook of International Corporate Governance.* Cheltenham: Edward Elgar.

Mansfield, N. R., Ugwu, O. O., and Doranl, T. (1994). Causes of Delay and Cost Overruns in Nigerian Construction Projects. *International Journal of Project Management*, 12(4), pp. 254–260.

Martínez, L. M. G., and Viegas, J. M. (2012). The Value Capture Potential of the Lisbon Subway. *Journal of Transport and Land Use*, 5(1), pp. 65–82.

Matos-Castano, J. (2011). *Impact of the Institutional Environment on the Development of Public Private Partnerships in the Road Sectors. Comparison of Two Settings: The Netherlands and Tamil Nadu.* PhD, University of Twente.

Matsukawa, T., and Habeck, O. (2007). Review of Risk Mitigation Instruments for Infrastructure Financing and Recent Trends and Developments, Trends and Policy Options No. 4. The International Bank for Reconstruction and Development, The World Bank, Washington, DC.

Miranda, N. (2007). Concession Agreements: From Private Contract to Public Policy. *The Yale Law Journal*, 117, pp. 510 – 549.

Mitusch, K., Pantelias, A., Vajdic, N., Syriopoulos, T., and Brambilla, M. (2015). Deliverable D2.3 – Financing Schemes Typology, BENEFIT (Business Models for Enhancing Funding and Enabling Financing for Infrastructure in Transport) Horizon 2020 Project, Grant Agreement No 635973. www.benefit4transport.eu/index.php/reports.

Mladenović, G., Roumboutsos, A., Campos, J., Cardenas, I., Cirilovic, J., Costa, J., González, M.M., Gouin, T., Hussain, O., Kapros, S., Karousos, I., Kleizen, B., Konstantinopoulos, E., Lukasiewicz, A., Macário, R., Manrique, C., Mikic, M., Moraiti, P., Moschouli, E., Nikolic, A., Nouaille, P.F., Pedro, M., Sintes, F.I., Soecipto, M., Trujillo Castellano, L., Vajdic, N., Vanelslander, T., Verhoest, C., and Voordijk, H. (2016). Deliverable D4.4- Effects of the Crisis & Recommendations, BENEFIT (Business Models for Enhancing Funding and Enabling Financing for Infrastructure in Transport) Horizon 2020 Project, Grant Agreement No. 635973. www.benefit4transport.eu/index.php/reports.

Morris, M., Schindehutte, M., and Allen, J. (2005). The Entrepreneur's Business Model: Toward a Unified Perspective. *Journal of Business Research*, 58, pp. 726–35.

Morris, S. (1990). Cost and Time Overruns in Public Sector Projects. *Economic and Political Weekly*, 15, pp.154–168.

Moschouli, E., and Vanelslander, T. (2016). Building a Typology of Modes of Transport with an Infrastructure Funding and Financing Perspective. In Venezia, E. (ed.), *Decisional Process for Infrastructural Investment Choice*, Transport. Heidelberg: Springer. In press.

Mota, J., and Moreira, A. C. (2015). The Importance of Non-Financial Determinants on Public Private Partnerships in Europe. *Journal of Project Management* 33, pp. 1563–1575.

Nijkamp, P., and Ubbels, B. (1999). How Reliable Are Estimates of Infrastructure Costs? A Comparative Analysis. *International Journal of Transport Economics*, 26(1), pp. 23–53.

Niles, J., and Nelson, D. (2001). Identifying Uncertainties in Forecasts of Travel Demand. *Proceedings Transportation Research Board 80th Annual Meeting*, Washington D.C.

Odeck, J. (2004). Cost Overruns in Road Construction – What Are Their Size and Determinants? *Transport Policy*, 11, pp. 43–53.

OECD. (2004). OECD Principles of Corporate Governance 2004.www.oecd.org (Accessed 17 February, 2015).

OECD. (2011). Strategic Transport Infrastructure Needs to 2030. OECD Futures Project on Transcontinental Infrastructure Needs to 2030/50, Paris.

OECD. (2014). OECD Statistics. http://stats.oecd.org/.

Onisko, A., Druzdzel, M.J., and Wasyluk, H. (2001). Learning Bayesian Network Parameters from Small Datasets: Application of Noisy-Or Gates. *International Journal of Approximate Reasoning*, 27(2), pp. 165–182.

Osei-Kyei, R., and Chan, A.P. (2015). Review of Studies on Critical Success Factors for Public Private Partnerships (PPP) Projects from 1990 to 2013. *International Journal of Project Management* 33, pp. 1335–1346.

Osterwalder, A. 2004. The Business Model Ontology – A Proposition in a Design Science Approach. Dissertation 173, University of Lausanne, Switzerland.

Painvain N. (2010). *High Speed Rail Projects: Large, Varied and Complex, Global Infrastructure & Project Finance*. Fitch Rating.

Percoco, M. (2014). Quality of Institutions and Private Participation in Transport Infrastructure Investment: Evidence from Developing Countries. *Transportation Research A*, 70, pp. 50–58.

Peters, B. G and Pierre, J. (2012). *The SAGE Handbook of Public Administration* (2nd edition). London: SAGE Publication

Pollitt, C. (2013). *Context in Public Policy and Managemen.*, Cheltenham: Edward Elgar.

Poppo, L., and Zenger, T. (2002). Do Formal Contracts and Relational Governance Function as Substitutes or Complements? *Strategic Management Journal*, 23(8), pp. 707–725.

Ragin, C. (2000). *Fuzzy-Set Social Science*. University of Chicago Press.

Ragin, C. C. (2008). *Redesigning Social Inquiry: Fuzzy Sets and Beyond*. London: University of Chicago Press, pp. 85- 94.

Reading, C. (2002). *Strategic Business Planning*. London: Kogan Page Limited.

Reeves, E. (2013). The Not So Good, the Bad and the Ugly: Over Twelve Years of PPP in Ireland. *Local Government Studies*, 39, pp. 375–395.

Roumboutsos, A., and Pantelias, A. (2015). Allocating Revenue Risk in Transport Infrastructure PPP Projects: How It Matters. *Transport Reviews*, 35(2), pp. 183–203.

Roumboutsos, A., Gouin, T., Leviäkangas P., Mladenović, G., Nouaille, P.-F., Voordijk, J., Moraiti, P., and Cardenas, I. (2016a). Transport Infrastructure Business Models: New Sources of Funding and Financing. *Proceedings of the WCTR 2016 Conference, The 14th World Conference on Transport Research*, Shanghai, July 10–14, 2016.

Roumboutsos, A., Bange, C., Bernardino, J., Campos, J., Cardenas, I., Carvalho, D., Cirilovic, J., González, M.M., Gouin, T., Hussain, O., Kapros, S., Karousos, I., Kleizen, B., Konstantinopoulos, E., Leviäkangas, P., Lukasiewicz, A., Macário, R., Manrique, C., Mikic, M., Mitusch, K., Mladenovic, G., Moraiti, P., Moschouli, E., Nouaille, P.F., Oliveira, M., Pantelias, A., Sintes, F.I., Soecipto, M., Trujillo, L., Vajdic, N., Vanelslander, T., Verhoest, K., Vieira, J., and Voordijk, H. (2016b). Deliverable D4.2 – Lessons Learned–2nd Stage Analysis, BENEFIT (Business Models for Enhancing Funding and Enabling Financing for Infrastructure in Transport) Horizon 2020 Project, Grant Agreement No. 635973. www.benefit4transport.eu/index.php/reports.

Roumboutsos, A., Pantelias, A., Sfakianakis, E., Edkins, A., Karousos, I., Konstantinopoulos, E., Leviäkangas, P., and Moraiti, P., (2016c). Deliverable D3.2- The Decision Matching Framework Policy Guiding Tool, Project Rating Methodology and Methodological Framework to Increase Business Model Creditworthiness, BENEFIT (Business Models for Enhancing Funding and Enabling Financing for Infrastructure in Transport) Horizon 2020 Project, Grant Agreement No. 635973. www.benefit4transport.eu/index.php/reports.

Roumboutsos, A., Bernardino, J., Brambilla, M., Carvalho, D., Campos, J., Cardenas, I., J., Duarte Costa, J., Gouin, T., Karousos, I., Leviäkangas, P., Liyanage, C., Lukasiewicz, A., Macário, R., Mikic, M., Mladenovic, G., Moraiti, P., Moschouli, E., Nouaille, P. F., Pantelias, A., Sintes, F. I., Trujillo Castellano, L., Vanelslander, T., Verhoest, K., and Voordijk, J.T., (2016d) Deliverable D5.3 – Policy Guidelines and Recommendations, BENEFIT (Business Models for Enhancing Funding and Enabling Financing for Infrastructure in Transport) Horizon 2020 Project, Grant Agreement No. 635973. www.benefit4transport.eu/index.php/reports.

Ruster, J. (1996). Mitigating Commercial Risks in Project Finance. Public Policy for the Private Sector, The World Bank, Note (69).

Saltelli, A., Ratto, M., Andres, T., Campolongo, F., Cariboni, J., Gatelli, D. Saisana, M., and Tarantola, S. (2008). *Global Sensitivity Analysis, The Primer*. John Wiley & Sons.

Scharfstein, D., and Stein, J. (2000). The Dark Side of Internal Capital Markets: Divisional Rent-Seeking and Inefficient Investment. *Journal of Finance*, 55, pp. 2537–2564.

Scharpf, F.W. (1991). Games Real Actors Could Play: The Challenge of Complexity. *Journal of Theoretical Politics*, 3, pp. 277–304.

Shafer, S. M., Smith, H. J., and Linder, J. (2005). The Power of Business Models. *Business Horizons*, 48, pp. 199–207.

Shenhar, A.J., and Dvir, D. (2007). *Reinventing Project Management*. Boston, MA: *Harvard Business School Publishing*, p. 5.

Shibani, A., and Arumugam, K. (2015). Avoiding Cost Overrun in Construction Projects in India. *Management Studies*, 3 (7–8), pp. 192–202.

Stevens, B., Schieb, P., and Andrieu, M. (2006). *Infrastructure to 2030: A Global Perspective on the Development of Global Infrastructures to 2030*. Paris: OECD.

Stewart, D. W., and Zhao, Q. (2000). Internet Marketing, Business Models and Public Policy. *Journal of Public Policy and Marketing*, 19, pp. 287–296.

Stowell, D. (2010). *An Introduction to Investment Banks, Hedge Funds, and Private Equity*. MA: Academic Press.

Susarla, A. (2012). Contractual Flexibility, Rent Seeking, and Renegotiation Design: An Empirical Analysis of Information Technology Outsourcing Contracts. *Management Science*, 58(7), pp. 1388–1407.

Tan, W. (2007). *Principles of Project and Infrastructure Finance*. Routledge.

Timmers, P. (1998). Business Models for Electronic Markets. *Electronic Markets*, 8(2), pp. 3–8.

Turner, J.R., and Keegan, A. (2001). Mechanisms of Governance in the Project-Based Organization: Roles of the Broker and Steward. *European Management Journal*, 19(3), pp. 254–67.

Ubbels, B., and Nijkamp, P. (2002). Unconventional Funding of Urban Public Transport. *Transportation Research Part D: Transport and Environment*, 7(5), pp. 317–329.

Van der Gaag, L. C. (1996). Bayesian Belief Networks: Odds and Ends. *The Computer Journal*, 39(2), pp. 97–113.

Vanelslander, T., Roumboutsos. A., Bernardino, J., Cardenas, I., Carvalho, D., Karousos, I., Moraiti, P., Moschouli, E., Verhoest, K., Voordijk, H., and Williems, T. (2015). Deliverable D2.2 – Funding Schemes and Business Models, BENEFIT (Business Models for Enhancing Funding and Enabling Financing for Infrastructure in Transport) Horizon 2020 Project, Grant Agreement No. 635973. www.benefit4transport.eu/index.php/reports.

Vasilescu, A.M., Dima, A. M., and Vasilache, S. (2009). Credit Analysis Policies in Construction Project Finance. *Management & Marketing*, 4(2), pp. 79–94.

Vassallo, J.M., and Baeza, M. (2007). Why Traffic Forecasts in PPP Contracts are Often Overestimated. EIB University Research Sponsorship Programme, EIB.

Vassalo, J., and Solino, A. (2006). Minimum Income Guarantee in Transportation Infrastructure Concessions in Chile. *Transport Research Record*. DOI:http://dx.doi.org/10.3141/1960-03.

Verhoest, K., Molenveld A., and Willems T. (2014). Explaining Self-Perceived Accountability of Regulatory Agencies in Comparative Perspective: How do Formal Independence and De Facto Managerial Autonomy Interact? Palgrave Macmillan, Series Title: Executive politics and governance, Host Document: *Accountability and Regulatory Governance*, pp. 51–77, Article number: 3, ISBN: 9781137349576, Public Governance Institute.

Verhoest, K., Peters, B.G., Bouckaert, G., and Verschuere, B. (2004). The Study of Organisational Autonomy: A Conceptual Review. *Public Administration Development*, 24, pp. 101–118.

Verhoest, K., Petersen, O.H., Scherrer, W., and Soecipto, R.M. (2015). How Do Governments Support the Development of Public-Private Partnerships? Measuring and Comparing PPP Governmental Support in 20 European Countries. *Transport Reviews*, 35(2), pp. 118–139.

Verhoest, K., Voets, J. and Van Gestel, K. (2012). A Theory-Driven Approach to PPP: The Dynamics of Complexity and Control. In Hodge, G. and Greve, C. (Eds.), *Rethinking Public–Private Partnerships: Strategies for Turbulent Times.* London: Routledge, pp. 188–210.

Voordijk, H., Roumboutsos, A., Cardenas, I., Łukasiewicz, A., Macário, R., Pantelias, A. (2015). Deliverable D2.4 – Governance Typology, BENEFIT (Business Models for ENhancing Funding and enabling Financing for Infrastructure in Transport) Horizon 2020 project, grant agreement No 635973. Available at http://www.benefit4transport.eu/index.php/reports.

Watson, D. (2005). *Business Models.* Petersfield: Harriman House.

Weill, P., and Vitale, M. R. (2001). *Place to Space: Migrating to E-Business Models.* Boston, MA: Harvard Business School Press.

White, L. (2001). Effective Governance through Complexity Thinking and Management Science. *Systems Research and Behavioural Science,* 18, pp. 241–57.

Williams, D., Marks, K. H., Robbins, L. E., Fernandez, G., and Funkhouser, J. P. (2009). *The Handbook of Financing Growth: Strategies, Capital Structure, and M&A Transactions.* New Jersey: Wiley Finance.

Williamson, O.E. (2000). The New Institutional Economics: Taking Stock, Looking Ahead. *Journal of Economic Literature,* 38(3), pp. 595–613.

Winch, G.M. (2001). Governing the Project Process: A Conceptual Framework. *Construction Management & Economics,* 19(8), pp. 799–808.

World Bank. (2014). Worldwide Governance Indicators. http://info.worldbank.org/governance/wgi/index.aspx#home

World Economic Forum (WEF). (2014). Global Competitiveness Report. www.weforum.org/reports/global-competitiveness-report-2014-2015.

Yescombe, E. R. (2007). *Public-Private Partnerships: Principles of Policy and Finance.* Butterworth-Heinemann.

Zagosdzon, B. (2013). Determinants of Implementation of Public Private Partnerships in Poland: The Case of Transport Infrastructure. *Advance Economic and Business,* 1(2), pp. 57–71.

Zhang, S., Zhang, S., Gao, Y., and Ding, X. (2016). Contractual Governance: Effects of Risk Allocation on Contractors' Cooperative Behavior in Construction Projects. *Journal of Construction Engineering and Management,* 142(6).

Zhao, Z.J., Vardhan, K.D., and Larson, K. (2012). Joint development as a value capture strategy for public transit finance. *The Journal of Transport and Land Use,* 5(1), pp 5–17.

Zott, C., Amit, R., and Massa, L. (2011). The Business Model: Recent Developments and Future Research. *Journal of Management,* 37(4), pp. 1019–1042.

4 Measuring transport infrastructure project resilience

Athena Roumboutsos and
Aristeidis Pantelias

4.1 Introduction

This chapter presents a new indicator that aims to assess the likelihood of a transport infrastructure project to achieve its target outcomes. The development of this indicator takes stock of findings from the qualitative and the indicator-based quantitative analyses that were presented in Chapters 2 and 3, respectively. The Transport Infrastructure Resilience Indicator (TIRESI), as its name suggests, aims to measure how resilient a project is, on its course to achieve its intended outcomes, in the face of its internal structure as well as its external conditions.

The methodological underpinnings for the TIRESI are based on a combination of scientific research and industry practice, as explained in the subsequent sections. It builds upon academic research in project and system resilience but also leverages elements from credit assessment methodologies that have been developed and applied by specialised firms in the financial services industry.

The TIRESI has three versions, namely a static (S-TIRESI), a dynamic (D-TIRESI) and an overall (O-TIRESI) version. Each version has a different role to play and its estimation refers to the following:

- The likelihood of reaching a specific outcome target at a particular point in the project's life cycle (S-TIRESI). This is the basic and most fundamental version of the indicator and the one that drives its implementation.
- The dynamic variation of the S-TIRESI with respect to external influences that could increase or diminish the likelihood of reaching a specific outcome target (D-TIRESI).
- The overall likelihood of attaining project outcome targets (either individually or in combination) as perceived by different project stakeholders (O-TIRESI).

The TIRESI takes values that are assessed based on a rating system that aims to categorise/classify projects based on the likelihood of achieving their target outcomes. This rating system is universal in terms of the possible rating categories, but getting to each one of them is not. The way different projects can end up belonging to each of the possible classifications is mode specific. This is because different modes present different sensitivities to changes of their internal and/or external conditions. In other words, project resilience for each mode has its own drivers. The quantification methodology of the TIRESI and the application of its rating system are, therefore, presented per mode of transport. This methodology provides the necessary tools for a full demonstration that is elaborated upon in Chapter 5 through the discussion of nine (9) cases of transport infrastructure projects from various modes and countries.

4.2 The Transport Infrastructure Resilience Indicator (TIRESI)

4.2.1 Project and system resilience: basic definitions and concepts

The term resilience (or resiliency) has its origin in the Latin term 'resilire', which means 'to leap/spring/bounce back; to recoil' (Merriam-Webster, n.d.; Oxford Dictionaries, n.d.). Resilience has been of interest to several disciplines such as ecology, sociology, psychology, organisational theory, engineering (e.g. networks, safety management, infrastructure systems, etc.) and economics. There is currently no consistent treatment of the underlying concept, with a tendency to use it as a 'buzzword' for various purposes and assessments.

Resilience has been associated with numerous concepts such as robustness, fault tolerance, flexibility, reliability, survivability, agility, redundancy, resourcefulness, rapidity, adaptability, absorptivity and recoverability, among others (Bruneau et al., 2003; Fisher et al., 2010; Henry and Ramirez-Marquez, 2012; Rose and Krausmann, 2013; Filippini and Silva, 2014; Francis and Bekera, 2014; Rose, 2015). It is sometimes also treated as the opposite of vulnerability although this is considered by many as an arbitrary assumption (Rose and Krausmann, 2013).

Research on resilience tends to fall within two major groups (Rose, 2015). The first group views resilience as any or all possible actions that can be put in place to reduce loss from a disruptive event (e.g. a disaster), ranging from pre-event mitigation to post-event recovery. The other group views resilience as actions following the materialisation of a disruptive event. Although the focus of the first group lies mostly on pre-event mitigation and is dominated by engineers, the second group acknowledges resilience as a process through which steps can be taken to enhance resilience even though some of them cannot be implemented until after the event has happened.

Resilience is recognised to be a time-dependent function related to the system's delivery function, that is the ability of the system to deliver its intended service (Henry and Ramirez-Marquez, 2012). This attribute of resilience makes necessary the consideration of its time path with respect to the recovery of the system and also creates the need to distinguish between short- and long-term recovery (Rose and Krausmann, 2013).

A few additional definitions and concepts are useful.

Resilience, with particular reference to economic resilience, is applicable at three distinct levels with respect to its impact and assessment (Rose and Krausmann, 2013):

- – micro-level: pertaining to individual businesses;
- – meso-level: pertaining to individual industries or markets;
- – macro-level: pertaining to the combination or all economic entities.

Resilience can be distinguished into static and dynamic. Static resilience is defined as the ability of a system to maintain its intended function when shocked. Dynamic resilience is defined as the ability of a system to expedite its recovery after experiencing a shock.

Resilience can also be distinguished as inherent or adaptive. Inherent resilience corresponds to aspects of recovery that are already built into the system. Adaptive resilience corresponds to actions and decisions undertaken under 'stress' that enhance the ability of the system to recover.

Resilience always has a cost associated with it and does not come for free. The cost of resilience actions is closely related to their corresponding effectiveness and should always be assessed with respect to it (Rose, 2015). At the same time, the total cost of the system is a combination of the cost from the implementation of resilience actions as well as the cost incurred due to the system's inability to deliver its intended function because of the disruptive event (Henry and Ramirez-Marquez, 2012).

4.2.2 Defining the Transport Infrastructure Resilience Indicator

Based on the above definitions and concepts, the TIRESI aims to quantify 'the ability of a Transport Infrastructure project to withstand, adjust to and recover from changes within its structural elements that affect its capability to deliver specific outcomes'.

Its 'stepping stones' are the elements of the BENEFIT Framework, represented through carefully selected operational indicators, and the understanding of their interrelations (see Chapter 3). As the proposed TIRESI indicator focuses on project implementation, it makes no differentiation between publicly financed and privately (co)financed projects. Its aim is to provide project stakeholders, whether experienced or not, with a measure of how likely it is that a particular project, under its specific conditions of delivery (expressed through indicator values of BENEFIT Framework elements) will be able to attain its project-specific goals in terms of cost and time to completion, as well as traffic and revenue forecasts. Furthermore, the TIRESI aims to assist in identifying which changes, whether internal or external, a project may *withstand, adjust to and recover from*.

The TIRESI has three different manifestations. These are defined as follows.

STATIC TRANSPORT INFRASTRUCTURE RESILIENCE INDICATOR

The S-TIRESI reflects the potential of reaching a specific project outcome target at a particular point in time. It is estimated based on the indicator values that are relevant to the respective outcome (cost to completion, time to completion, actual vs. forecast traffic and actual vs. forecast revenue). This is the basic and most fundamental version of the indicator and the one that drives its implementation.

DYNAMIC TRANSPORT INFRASTRUCTURE RESILIENCE INDICATOR

The D-TIRESI is assessed based on the S-TIRESI and represents the vulnerability (or stability) of the S-TIRESI with respect to the outcome target under consideration.

OVERALL TRANSPORT INFRASTRUCTURE RESILIENCE INDICATOR

Finally, the O-TIRESI is an expression of all outcome ratings, which may be based on the static or dynamic expression of the indicator.

The development and quantification of the various versions of the TIRESI capitalises on existing scientific work as well as professional practice. It leverages methodological elements from resilience models that have been published in the academic literature as well as credit assessment methodologies that reflect existing market practice. Quantifying the S-TIRESI constitutes the core of the methodology as both D-TIRESI and O-TIRESI are based on it. The principles of its quantification, in terms of methodological considerations, are described in Section 4.2.3. Its full operationalisation is described per outcome and infrastructure mode in Section 4.4.

4.2.3 Measures of resilience and methodological considerations

Various measures of resilience currently exist in the literature and there is no consistent quantitative approach for its assessment. These measures differ depending on the underlying definition that is used as well as the system under consideration. A wide range of different approaches currently exists, many of which can be useful in the context of transport infrastructure funding and financing. Indicatively, such measures can be perused in Rose and Krausmann (2013), Henry and Ramirez-Marquez (2012), Filippini and Silva (2014) and Fisher *et al.* (2010), among many other relevant studies.

Additionally, although not directly branded as such, credit assessment methodologies aim to treat a very specific aspect of infrastructure project resilience. These methodologies evaluate the likelihood of a project to repay its obligations against debt financing received from capital market financiers. Various versions of these methodologies exist as they tend to be proprietary to their developers as well as sector/industry specific. Their particular characteristics and relevance to project stakeholders are briefly discussed in Chapter 6.

Through the consideration of the above-mentioned measures of resilience, several modelling principles as well as good practices were identified. A few have been considered particularly relevant and important in the context of transport infrastructure project resilience and have formed the methodological considerations behind the development and quantification of the TIRESI.

CONSIDERATION 1: SYSTEM BOUNDARIES

The first consideration is the need to clearly identify the limits of the system of interest (Henry and Ramirez-Marquez, 2012). Drawing the system boundary may not be a trivial task, especially in complex systems. At the same time, this boundary helps distinguish between internal and external disruptive events that may affect the system.

The TIRESI considers the system nature of the BENEFIT Framework, which communicates with the wider universe of infrastructure delivery through its inputs, that is all decisions made prior to project award. These decisions include the project budget, the expected construction duration as well as forecast traffic and revenues. The implementation of the project, which is the focus of the BENEFIT Framework, is represented through indicators. Among these indicators, two of them, the Financial-Economic Indicator (FEI) and the Institutional Indicator (InI), describing the implementation context, are considered exogenous to the project system. All other indicators are considered endogenous.

CONSIDERATION 2: FIGURE OF MERIT

The second consideration is the specification of the system's figure-of-merit[1] (Henry and Ramirez-Marquez, 2012). This contributes to the quantification of resilience as a function of the system's ability to deliver its intended service. It is possible that a system may have multiple figures of merit as well as that it may exhibit resilience for one figure of merit but not for another.

While the outcomes of a transport infrastructure project can be many, the BENEFIT Framework has been operationalised by focusing on four: cost to completion, time to completion, actual versus forecast traffic, and actual versus forecast revenue. These four outcomes correspond to the four different figures of merit of the TIRESI. As expected, the system may exhibit simultaneous resilience for one or more figures of merit (outcomes), but not necessarily for all.

CONSIDERATION 3: ACTIONABLE VARIABLES AND BACKGROUND CONDITIONS

The third consideration is that introducing the concept of resilience should not just aim to study the relevant recovery process, but also to improve it (Rose and Krausmann, 2013). In that respect, it needs to distinguish between actionable variables and background conditions. The former are the ones that can be tied to decision making, while the latter can be considered as non-crucial for the recovery process as they cannot be influenced.

The TIRESI is structured by carefully considering the distinct nature and impact of the identified endogenous and exogenous indicators (see Chapter 3). In the developed methodology, it becomes immediately obvious which indicators need to be addressed to improve resilience (actionable indicators) and which are simply considered as background conditions. Notably, the exogenous indicators are non-actionable, while the endogenous provide varying levels of flexibility to decision makers. The latter are also distinguished into structural and policy indicators. Structural indicators are set at the early stages of project delivery and require significant effort to change, while policy indicators allow for greater flexibility (under conditions) throughout the life time of the project. For example, the Governance Indicator (GI) is closely connected to the contractual conditions. Once a contract is signed, changing basic contractual clauses is not always easy or advisable. In contrast, changing toll rates on a motorway is more common. However, this may come with a change in the risk profile of the revenue and/or remuneration scheme influencing the value of the Revenue Robustness Indicator (RRI) or the Remuneration Attractiveness Indicator (RAI), or both.

CONSIDERATION 4: SYSTEM STABILITY

The fourth consideration is that in the face of a disruptive event, the recovery of the system will be dictated by two sets of parameters: those that pertain to the nature of the system and those that pertain to pre-determined policies and available facilities for recovery (Henry and Ramirez-Marquez, 2012). This consideration matches the concept of inherent resilience (Rose and Krausmann, 2013), but underplays the significance of adaptive resilience as it puts the emphasis on the system. Adaptive resilience appears to be more related to the capabilities of decision makers and, in that respect, should also have an important role to play as their decisions 'under stress' could influence profoundly the way and speed by which the system will bounce back from its disrupted state.

It is acknowledged that the transport infrastructure delivery and operation system (BENEFIT Framework) and its representation through indicators will change over time, especially as a response to involuntary changes in the exogenous indicators. In this context, it is not enough to provide a TIRESI value reflecting current conditions (S-TIRESI). This indicator needs to be accompanied by an indication of potential resilience to change. To this end, the methodology also defines a D-TIRESI.

CONSIDERATION 5: ACCURACY AND TRANSPARENCY

The fifth and final consideration is that any resilience measure introduced needs to be based on information that is accurate and transparent (Fisher *et al.*, 2010). Of particular significance is also the reproducibility of the measure as its value and usefulness is closely related to the ability to compare it and interpret it in a consistent way.

The TIRESI methodology is systematic, consistent and does not require a qualitative assessment or interpretations. Therefore, the resulting TIRESI values are both reproducible and easy to verify.

4.2.4 *Transport Infrastructure Resilience Indicator rating system*

In Chapter 3, different values and combinations of indicators were found to drive projects towards the attainment of specific outcomes. The TIRESI takes stock of these combinations of indicators and their respective values in its quantification methodology. It then uses this information to classify/categorise projects based on the resulting likelihood of achieving their outcomes. This classification/ categorisation is done using a novel rating system, which has been developed for this specific purpose and comprises three basic rating categories, namely A, B and C. The proposed rating categories are specified as follows:

Rating A: Describes very high likelihood of reaching the figure of merit target value (achievement of outcome).

Projects assigned an A rating exhibit high values of exogenous (FEI and InI) indicators and combinations of endogenous (all other) indicator values that secure the attainment of the specific target.

Rating B: Describes average likelihood of reaching the figure of merit target value (achievement of outcome).

A project assigned a B rating exhibits potential vulnerability that may be due to either exogenous (FEI and InI) or endogenous (all other indicators) conditions. Because of these two different sources of vulnerability, this rating category is further divided into B_{EX} and B_{EN}, corresponding to:

- B_{EX}: A rating describing a fairly robust internal project structure but subject to exogenous vulnerability; and
- B_{EN}: A rating describing a project implemented under largely positive exogenous conditions but with internal structure vulnerabilities.

Rating C: Describes low likelihood of reaching the figure of merit target value (achievement of outcome).

Projects assigned a C rating are vulnerable to both exogenous and endogenous conditions or the respective exogenous (endogenous) conditions are not sufficient to balance the endogenous (exogenous).

Furthermore, due to the large number of indicators involved in determining each rating for each figure of merit and transport mode, and the resulting large number of their corresponding combinations, slightly better or worse conditions may exist. These are presented with additional rating notches, (+) or (−), shown next to the basic rating, A, B or C. Table 4.1 presents the full range of potential classifications/categories of the proposed rating system.

Table 4.1 Transport Infrastructure Resilience Indicator rating system

Exogenous vulnerability	Rating category	Endogenous vulnerability
None	A^+	None
None	A	Limited
None	A_-	Some
Some	B_{EX}	Limited
Endogenous structure reduces vulnerability	B_{EX}^+	Limited
Endogenous structure increases vulnerability	B_{EX-}	Limited
Limited	B_{EN}	Some
Limited	B_{EN}^+	The combination of endogenous and exogenous conditions reduces vulnerability
Limited	B_{EN-}	The combination of endogenous and exogenous conditions increases vulnerability
Existing: the combination of endogenous and exogenous conditions reduces vulnerability	C^+	Existing: the combination of endogenous and exogenous conditions reduces vulnerability
Existing	C	Existing
Significant	C_-	Significant

Source: Authors adjusted from Roumboutsos *et al.* (2016).

4.3 Drivers of transport infrastructure resilience

Chapter 2 concluded that the elements of the BENEFIT Framework influence project outcomes in each transport mode albeit with different factors having a different impact on each. In Chapter 3, expressing these key elements of the 'transport infrastructure delivery and operation/maintenance system' (BENEFIT Framework) through quantifiable indicators enabled a deeper and more extensive analysis of the projects of the BENEFIT project case dataset. Through the comparison of results from the multi-method analysis approach followed, over-arching observations were made with respect to the combinations of indicators that drive project outcomes. However, a different number of project cases exists per mode of transport in the BENEFIT project case dataset. In this context, the findings from Chapter 3 need to be carefully scrutinised by considering also the results from the qualitative analysis per mode that was presented in Chapter 2.

In the present section, the importance of different indicators in the attainment of specific project outcomes is assessed for each mode following a brief description of the methodological approach adopted.

4.3.1 Methodological approach

A heuristic approach was followed in identifying the indicators driving outcomes per transport mode. The process followed is briefly outlined herewith (see Roumboutsos *et al.*, 2016, for a detailed description).

The starting point of the heuristic process was the findings and conclusions from the comparison of results from the multi-method analysis of indicators describing the projects of the BENEFIT project case dataset (Chapter 3). The process then comprised the formulation of hypotheses with respect to possible indicator combinations driving outcomes for each mode by considering:

- the composition of the BENEFIT project case dataset with respect to the representation of different infrastructure modes and countries of project origin;
- the individual results of each indicator analysis method (Chapter 3). For example, the fsQCA identifies 'paths' (i.e. combinations of indicators leading to/or not to a specific outcome) and for each 'path' provides the set of project cases that totally follow this 'path' (membership in set) or not (non-membership in set). Similarly, the econometric analysis provides information on marginal effects, etc.;
- the qualitative review of the BENEFIT project case dataset, and other relevant research reported in the literature (Chapter 2).

The relevant subsamples of the BENEFIT project case dataset, were, then, scrutinised with respect to the formulated hypotheses. Cases that did not conform to the hypotheses were investigated further to identify the underlying reasons through:

- causal mapping and
- in-depth qualitative analysis.

The investigation of noncomplying project cases led either to confirm that the reasons of noncompliance were related to inappropriately set outcome targets (e.g. overestimated traffic forecasts) or to knowledgeably revisit the formulation of the original hypotheses.

This process was repeated for each outcome and each transport mode until all project cases could be explained.

Figure 4.1 illustrates the methodological process followed.

4.3.2 Road infrastructure

Across all road projects in the BENEFIT project case dataset, the FEI appears to be influential with respect to traffic outcomes but has also an impact on cost and

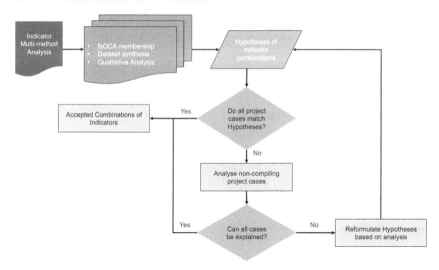

Figure 4.1 Methodological approach in identifying drivers (indicators) of infrastructure resilience per mode

time targets. A high InI may limit the impact of a low FEI. In this context, countries with high InIs are more capable of weathering a financial/economic crisis.

With respect to time to completion, while the GI is the one dominating the potential to achieve this outcome, a high CSI may compensate smaller GI values. Trade-offs between cost and time to completion have been found to take place. In PPPs, there is also a trade-off with respect to revenues and traffic, especially if high-risk remuneration schemes are in place.

Table 4.2 summarises indicators influencing road project outcomes.
Key findings may be summarised as follows:

Indicators exogenous to the project

- The FEI plays a significant role in road projects. An increase or decrease in the value of this indicator may have a respective impact on the probability of achieving time to completion, cost to completion and traffic targets. The influence is greater with respect to traffic and smaller with respect to time to completion. A high value of the InI may off-set the impact of a low FEI on time targets. The FEI is not a determining indicator with respect to the revenue target.
- The InI may be the most important. While exogenous to the project, it is not affected by economic cycles and, therefore, describes a measure of resilience to financial shocks. It is a pre-requisite in achieving cost and time to completion targets, while a high value may also limit the impact of a low FEI on traffic. Once again, the InI is not a determining indicator with respect to the revenue target.

Table 4.2 Summary table of indicators influencing outcomes in road projects

Indicators	Outcomes			
	Cost to completion	Time to completion	Actual vs forecast traffic	Actual vs forecast revenue
Financial – Economic Indicator (FEI)	Strong positive or negative influence depending on high or low value	Positive or negative influence depending on high or low value (May be off-set by GI and InI)	Very strong Positive or negative influence depending on high or low value	
Institutional Indicator (InI)	Pre-requisite	Pre-requisite (Acts in combination with GI)	High value may limit effect of FEI	
Governance Indicator (GI)	Needed (compensates for low CSI)	Pre-requisite (acts in combination with InI)	High value may limit effect of FEI	Support: high value
Cost Saving Indicator (CSI)	Needed (compensates for low GI)	Needed	High value may limit effect of FEI	Support: high value
Revenue Support Indicator (RSI)	Support			Expected for high value
Remuneration Attractiveness Indicator (RAI)		Driver: Low values	High value may limit effect of FEI	Support: high value
Revenue Robustness Indicator (RRI)		Driver: Low values		Key indicator
Financing Scheme Indicator (FSI)	Driver: high values			Expected for high value

Source: Authors adjusted from Roumboutsos *et al.* (2016).

Indicators endogenous to the project

• The GI reflects in many ways the level of institutional maturity in the country where the project is procured. In this respect, it may compensate and/or enhance the InI. As it describes the institutional arrangements within the project, it practically influences all project outcomes.

- The CSI describes the project's technical difficulty and also the capabilities of key project actors: the builder's to construct, the operator's to operate, and the monitoring authority's to monitor the project in consideration. In addition, it assesses whether capabilities are aligned with the risk allocation among these actors. All these attributes were found to be important throughout the project life cycle. A high CSI may compensate for a lower value of GI with respect to cost to completion and may also contribute to off-setting the impact of a low FEI on traffic.
- The importance of the Revenue Support Indicator (RSI) is limited in road projects and depends on the criticality and exclusivity of the project in the network. A high value of the factor 'Level of Coopetition' may have a positive impact on cost to completion and revenue targets.
- The RAI practically acts as a policy tool. Low values of the indicator will drive the attainment of time to completion targets. High values of the indicator will limit the effect of FEI on traffic and also support revenue targets.
- The RRI expresses the riskiness of the project revenue streams as well as the estimated level of cost coverage. Therefore, it becomes a key indicator in assessing the potential of achieving revenue targets and also drives the project towards being 'on-time'.
- The FSI is also a policy tool as it becomes a response to adverse exogenous indicators. Low values of the FEI and InI dictate the need for higher values of the FSI. In other words, countries with low FEI and InI values are forced to increase public contributions to project financing structures or opt out of the PPP model for project delivery altogether.

4.3.3 Urban transit infrastructure

What is noticeable in the delivery of urban transit projects is that to secure positive outcomes, all indicators identified have to exhibit high values. This is particularly important to achieve the cost to completion target. Some flexibility was identified with respect to the combination of the influence of the CSI and GI, as the high value of one indicator could compensate for a lower value in the other.

For these projects, the IRA was also identified as significant in achieving ridership and revenue targets. In Table 4.3 below, it is noted as a prerequisite.

Key findings may be summarised as follows:

Indicators exogenous to the project

- The FEI, whether describing the national context, as in the BENEFIT Framework, or the local conditions, is not decisive for urban transit projects. The effect of a low or decreasing FEI (recession) may affect ridership positively or negatively (or both). The only definite negative effect FEI may have, is on the amount of commercial revenues (e.g. advertisements), which is a common income stream for urban transit.

Table 4.3 Summary table of indicators influencing outcomes in urban transit projects

Indicators	Outcomes			
	Cost to completion	Time to completion	Actual vs forecast traffic	Actual vs forecast revenues
Institutional (InI)	High value	High value	High value	High value
Financial Economic (FEI)				Only with respect to advertisements
Governance (GI)	High value	High value (may be combined with CSI)	High value (may be combined with CSI)	High value
Cost Saving (CSI)	High value	High value (may be combined with GI)	High value (may be combined with GI)	High value
Revenue Support (RSI)	High value	High value	High value (with emphasis on LoC)	High value
Remuneration Attractiveness (RAI)			Support	High value
Revenue Robustness (RRI)				High value
Reliability/ Availability (IRA)			Prerequisite	Prerequisite
Financing Scheme (FSI)				
Comment	All indicators above should have high values			At least two of the above indicators should bear a high value

Source: Authors adjusted from Roumboutsos *et al.* (2016).

- The InI may be the most important. While exogenous to the project, it is not affected by economic cycles and, therefore, describes a measure of resilience to financial shocks as well as transparency with respect to project maturity. It was identified as a prerequisite for achieving all target outcomes.

Indicators endogenous to the project

- The GI reflects in many ways the level of institutional maturity in the country of project procurement. In this effect, it may compensate and/or enhance the InI. A high value contributes to the cost to completion and revenue targets. High values may also compensate for low values of the CSI for 'on-time' and traffic targets.
- A higher value of the CSI may compensate for a lower value of GI with respect to time to completion and traffic targets.
- The RSI is highly important in urban transit. It describes both the exclusivity of the project in the urban transit network as well as its ability to generate revenues from other sources. However, within urban transit, the most important factor is 'level of coopetition'.
- The RAI is always relatively high in urban transit as most operations are subsidised. Higher values of the indicator correspond to greater public support and consequently less expensive fares driving both ridership and revenues for the project.
- The RRI expresses the riskiness of the project revenue streams as well as the cost coverage level estimated. Therefore, it becomes a key indicator in assessing the potential of achieving revenue targets.
- The FSI was not found to have a particular impact on outcomes most probably because urban transit is usually heavily supported by the public sector. This support is mostly reflected in the RAI.
- Finally, while a high IRA is important for all modes, it is a prerequisite for traffic and revenue targets in urban transit projects.

4.3.4 Bridge and tunnel infrastructure

In bridge and tunnel projects, the FEI influences traffic and revenues but to a lower extent than roads due to the high level of coopetition that characterises these structures. The influence of the GFC on cost and time to completion could not be assessed, as all projects in the BENEFIT project case database were completed before the crisis. It is reasonable to assume that indicators applicable to road projects are of equal importance during the construction phase of bridge and tunnel projects.

Once again, the CSI, GI and RSI are important for all outcomes, with greater emphasis on the RSI (including the level of coopetition), as bridge and tunnel projects are usually designed with a low value of the RAI (example: user tolls).

Finally, the findings with respect to bridge and tunnel projects should be considered purely indicative given their small sample size in the BENEFIT project case database. These are summarised in Table 4.4.

Table 4.4 Summary of indicators influencing outcomes in bridge and tunnel projects

Indicators	Outcomes			
	Cost to completion	Time to completion	Actual vs forecast traffic	Actual vs forecast revenues
Institutional (InI)	High value	High value	High value (prerequisite for low RAI)	High Value (prerequisite for low RAI)
Financial Economic (FEI)	High value important. Low values may be off-set by high values of the other indicators			
Governance (GI)	High value	High value	High value (prerequisite for low RAI)	High value (prerequisite for low RAI)
Cost Saving (CSI)	High value	High value	High value (prerequisite for low RAI)	High value (prerequisite for low RAI)
Revenue Support (RSI)	High value (high LoC important)	High value (high LoC important)	High value (high LoC important)	High value (high LoC important)
Remuneration Attractiveness (RAI)		Low value (may compensate for RRI)		
Revenue Robustness (RRI)		Low value (may compensate for RAI)		High value
Reliability/ Availability (IRA)				
Financing Scheme (FSI)	High value	High value	High value	High value

Source: Authors adjusted from Roumboutsos et al. (2016).

Key findings may be summarised as follows:

Indicators exogenous to the project

- The FEI is an important indicator, as in the case of road projects. FEI will affect cost and time to completion targets and a high FEI is important in the case of a low RAI, which is common in bridge and tunnel projects. However, in the case of bridge and tunnel projects, the negative impact of a low and/or decreasing FEI may be offset by high values of other indicators (e.g. GI, CSI and RSI).

- The InI may be, once again, the most important. It is a prerequisite in achieving traffic and revenue targets in the case of a low RAI.

Indicators endogenous to the project

- A high value of the GI contributes to the 'on-budget'" and 'on-time' targets. High values are a prerequisite in achieving traffic and revenue targets in the case of a low RAI.
- A high value of the CSI contributes to the successful attainment of 'on-budget' and 'on-time' targets. High values are a prerequisite in achieving traffic and revenue targets in the case of a low RAI.
- The RSI is highly important for bridge and tunnel projects, especially its factor 'level of coopetition'. It describes the exclusivity of the project in the network and a high value has a positive impact on all outcomes.
- The RAI is typically low in bridge and tunnel projects as the relative high 'level of coopetition' allows the public sector to pass demand risk to the operator. A low RAI is a driver towards achieving time to completion. The RRI has the same effect with respect to achieving time targets. A low RAI may counterbalance a high RRI and vice versa.
- The RRI for bridge and tunnel projects has the same effect as RAI, that is it drives time to completion.
- The FSI was typically found to be high in all bridge and tunnel cases of the BENEFIT project case dataset. A high value of the FSI contributes to achieving project outcome targets.

In conclusion, it was identified that Bridges and Tunnels could be considered as special cases of road projects with much higher exclusivity, which allows them to attain low RAI values (high risk remuneration schemes such as tolls). As these are usually projects of high technical difficulty, their ability to reach construction phase objectives is dependent, in addition to design maturity, on the expertise of the contractors/concessioners.

4.3.5 Other transport infrastructure

The same methodology was applied to transport infrastructure projects concerning airports, ports and rail. However, the small sample size of these infrastructure modes in the BENEFIT project case dataset limited the explanatory power of the analysis as well as the ability to generalise conclusions. Therefore, for these transport modes, the presented resilience drivers may only be considered indicative.

4.3.5.1 *Airport infrastructure*

As in the case of other transport infrastructure, the CSI and GI indicators are important in achieving cost to completion targets. RSI also seems to be contributing positively. The same indicators are also important to achieve time to completion targets. Traffic is influenced by the implementation context (FEI, InI) but also by the way the airport is connected to the local/regional or national economy. Incentives connected to demand risk allocation are also important.

Airport revenue targets are not directly impacted by the FEI. For this outcome, RSI, RRI, CSI and GI are the indicators that drive positive results. Notably, the RAI is typically low in airport projects.

4.3.5.2 *Port infrastructure*

The prominent feature of the port cases is their uniqueness with respect to outcomes. Required conditions in achieving cost and time to completion include high values of the CSI, GI, RSI and InI indicators. However, very importantly, positive outcomes may be reached even if these conditions are not met.

Strategic behaviour from port operators seems to influence traffic and revenue outcomes. In effect, port traffic is dependent on the international strategies of shipping lines and hinterland connections (logistic supply chains), and, in this context, the level of coopetition factor, expressing the uniqueness of the port in the internal logistic networks that it serves, is of dominant importance.

Finally, the FEI continues to be important albeit in an extended context, which encompasses the logistic chains the port serves.

4.3.5.3 *Rail infrastructure*

With respect to their construction, rail projects have a similar (or even greater) technical risk profile to special structures (bridge and tunnel projects). However, contrary to special structures, they are not characterised by exclusivity and, therefore, cannot take advantage of their position in the network, as bridge and tunnels projects do. Consequently, rail projects usually exhibit higher values of the RAI.

In addition, while their RSI with respect to position in the network is relatively low (low exclusivity), other features can provide a high overall RSI such as business orientation, as well as other transport related (and non-transport-related) services. Notably, the bundling of services, while positive for revenue prospects, places pressure on both the GI and the CSI with respect to the capability of the contracting authority. The InI is important, also, in this context. Finally, rail projects usually exhibit high values of the FSI, as they require considerable support from the public budget.

Stemming from the above, while the initial tendency would be to approximate rail infrastructure projects to other linear infrastructures (such as roads), their behaviour with respect to achieving their outcomes makes them more similar to urban transit projects.

4.4 Transport Infrastructure Resilience Indicator quantification methodology

4.4.1 Overarching concept

As presented in Chapter 3, the indicators representing the elements of the BENEFIT Framework may be divided in two broad categories. These categories reflect the degree of influence a decision maker may have on them and, in turn, on the project's outcomes. The two categories are:

- Exogenous indicators: these include the FEI and the InI. These indicators describe the implementation context. The decision maker has no influence over these indicators and their values.
- Endogenous indicators: All other indicators fall under this category. This category is also divided into the following two groups:

  - Structural indicators: these describe the business model and the contractual conditions of implementation (IRA, CSI, RSI and GI). Notably, following project award, the flexibility and, therefore, the range of possible available decisions gradually becomes limited.
  - Policy indicators: the values of these indicators (RAI, RRI and FSI) may be changed (under specific terms) throughout the life cycle of the project based on corresponding decisions. Policy indicators, in combination with other indicators and for specific values of all the indicators combined, have the ability to induce incentives and drive particular aspects of project performance.

The relations between the indicators and the categories they belong to are illustrated schematically in Figure 4.2. This formulation, coupled with an understanding of the drivers of resilience identified for each transport mode, form the overarching concept of the TIRESI quantification methodology.

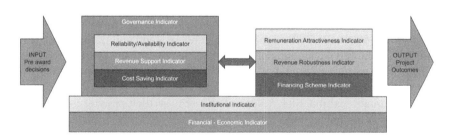

Figure 4.2 BENEFIT Framework: figurative relation between indicators

4.4.2 Static Transport Infrastructure Resilience Indicator (S-TIRESI) quantification methodology

Having identified the drivers of transport infrastructure resilience (i.e. combinations of indicators) for each transport mode, the next important milestone is to determine their respective threshold values, based on which projects will be rated. The approach includes four steps. In the first, general threshold values that could apply across all infrastructure modes are selected for all indicators. In a second instance, these general thresholds are calibrated against the BENEFIT project case dataset to adjust their values to each infrastructure mode and outcome. Then, in the third step, capitalising on the findings of Section 4.3 (resilience drivers per mode and outcome) and by considering the results from Step 2, projects in the BENEFIT project case dataset are assigned a rating. These ratings are based on the categories described in Section 4.2. Through the repetition of this process, a multi-dimensional space of rating 'trees' describing the possible combination of indicators related to a specific outcome and mode is generated in the fourth and final step. This multi-dimensional indicator space forms the basis for the assessment of the TIRESI.

The details of each step of the four-step process are presented in the following sections.

4.4.2.1 Determining general threshold values of resilience drivers (Step 1)

In the previous mode-specific discussion (Section 4.3), reference is made to the 'high' or 'low' value of different indicators. As explained in various instances, the reference to 'high' or 'low' values corresponds to different thresholds for each indicator that determine its influence on a specific outcome (figure of merit), within the context of a specific infrastructure mode. The typical 'high' and 'low' threshold values of each indicator, per outcome and mode, were determined (and calibrated) through relevant research. Two approaches were considered, which were subsequently combined.

FUZZY-SET QUALITATIVE COMPARATIVE ANALYSIS

As mentioned in Chapter 3, the fsQCA was one of the methods applied for the analysis of indicators. The fsQCA employs a direct calibration of the variables used (Ragin, 2008). The method uses estimates of the log of the odds of 'full membership' of a variable-(indicator) in a 'path', and considers three important qualitative anchors to structure the calibration. The resulting thresholds are defined as (Ragin, 2000): the threshold for full membership, the threshold for full non-membership and the cross-over point. Determined by following a rigorous approach (for more details see Roumboutsos *et al.*, 2016, pp. 95–119), the resulting threshold values for all indicators are listed in Table 4.5.

Table 4.5 Indicator threshold values

Indicators	Indicator value range		BENEFIT project cases indicator range		Threshold values
	Min	Max	Min	Max	
Institutional (InI)	0.000	1.000	0.380	0.860	0.650
Financial Economic (FEI)	0.000	1.000	0.305	0.842	0.600
Governance (GI)	0.000	1.000	0.188	0.875	0.600
Cost Saving (CSI)	−0.333	1.000	−0.300	1.000	0.333
Revenue Support (RSI)	0.000	1.000	0.000	0.416	0.500
Remuneration Attractiveness (RAI)	0.000	1.000	0.000	1.000	0.500
Revenue Robustness (RRI)	0.000	1.000	0.000	1.000	0.500
Reliability/ Availability (RAI)	0.000	1.000	0.250	1.000	1.000
Financing Scheme (FSI)	0.000	1.000	0.000	1.000	0. 650

Source: Roumboutsos et al. (2016).

FINANCIAL ECONOMIC INDICATOR: SECOND LOWER THRESHOLD

The financial economic context is known to influence demand in transport. The indicator analysis (Chapter 3) also highlighted the importance of the FEI in achieving outcomes. In effect, marginal changes in the FEI increase (or reduce) the probability of achieving outcomes by a significant probability (e.g. 58 and 72 per cent for cost to completion and actual vs. forecasted traffic, respectively). Observations of the value of FEI associated with lower performance indicated that for FEI < 0.50, projects, across all modes and outcomes, had a lower probability of reaching their respective targets.

4.4.2.2 Calibration of threshold values of resilience drivers (Step 2)

The threshold values determined in Step 1 were, then, calibrated for each mode. Special consideration was placed with respect to the Revenue Support Indicator, whose range of possible values differs between modes.

REVENUE SUPPORT INDICATOR: VALUE RANGE PER TRANSPORT INFRASTRUCTURE
MODE – MAXIMUM VALUES

For the composite indicator, RSI, it is important to note that its maximum value (RSI = 1) cannot be achieved by all transport modes and, therefore the threshold values are different per mode. Considering the structure of the RSI (see Chapter 3, Section 3.3.1), the theoretical maximum values of the indicator for each mode

Table 4.6 Revenue Support Indicator range and threshold values per mode

Mode	Indicator value range		BENEFIT cases indicator range		Threshold values
	Min	Max	Min	Max	
Roads	0.000	0.400	0.042	0.301	0.150
Urban transit	0.000	0.933	0.045	0.274	0.400
Bridges and tunnels	0.000	0.533	0.107	0.416	0.250
Airports	0.000	0.933	0.000	0.402	0.400
Ports	0.000	0.800	0.210	0.400	0.350

Source: Roumboutsos *et al.* (2016).

were identified. These values were then considered in the fsQCA calibration approach. The results are summarised in Table 4.6.

THRESHOLD VALUES FOR OTHER INDICATORS

In order to define the threshold values for all other endogenous indicators, a heuristic iterative approach was followed based on:

- the conclusions of Chapter 3;
- the identified drivers (combinations of indicators) per infrastructure mode and outcome (see Section 4.3);
- the BENEFIT project case dataset.

A similar process of iterations, as described in Section 4.3, was followed. Notably, considering the approach followed in determining threshold values, these should be considered as fuzzy rather than crisp.

4.4.2.3 *Generating the first cells of the rating space (Step 3)*

Based on Steps 1 and 2, as described previously, all projects in the BENEFIT project case dataset were assigned to rating categories with respect to the four figures of merit (outcomes) considered. Depending on the transport mode they belonged to, project-specific ratings capitalised on the findings from the analysis of mode-specific resilience drivers that were presented in Section 4.3.

For example, in the figure of merit 'cost to completion' for road infrastructure (Table 4.7):

- Lower values of the CSI and RSI may be compensated by higher values of the GI along with higher values of the InI.
- Lower values of the InI (but not lower than 0.60) may be compensated by a FSI with a value greater than 0.60. However, if the FSI is very small

(<0.333), then the prevailing strategy is to trade-off cost to completion for time to completion and a cost overrun is to be expected. This is reflected in the resulting rating 'C'.

Table 4.7 describes the rating assessment for the figure of merit 'cost to completion' for road infrastructure projects, which was used for the rating of all road projects in the BENEFIT project case dataset.

Table 4.7 Transport Infrastructure Resilience Indicator rating cost to completion for road infrastructure projects

	FEI	InI	GI	CSI	RSI	FSI
Max resilience						
Rating: A						
A_- for InI $\in$ [0.61, 0.65] and FSI > 0.60	≥0.60	≥0.65	≥0.500	≥0.333	≥0.150	
Endogenous vulnerability						
Rating: B_{EN}						
$B_{EN}{}^+$ for larger values of GI, CSI & RSI	≥0.60	≥0.65	≥0.700	[0.333, 0.000]	[0.150, 0.000]	
B_{EN-} for smaller values of GI						
B_{EN-} for InI $\in$ [0.61, 0.65] and FSI > 0.60						
Exogenous vulnerability						
Rating: B_{EX}						
$B_{EX}{}^+$ for larger values of GI. CSI & RSI	[0.50, 0.60]	≥0.65	≥0.500	≥0.333	≥0.150	≥0.600
B_{EX-} for smaller values of CSI & RSI						
B_{EX-} for InI $\in$ [0.61, 0.65] and FSI > 0.60						
B_{EX-} when C^+ and FSI > 0.666						
Poor resilience						
Rating: C						
C^+ for larger values of GI	<0.50	<0.65	<0.500	<0.333	<0.150	<0.60

Source: Roumboutsos *et al.* (2016).

This process was repeated for all figures of merit and all modes for all project cases. The full description of the assessment of ratings per figure of merit (outcome) and infrastructure mode may be found in Roumboutsos *et al.* (2016).

4.4.2.4 *Developing the multi-dimensional rating space: generation of rating trees (Step 4)*

The starting point for this activity is the 'cells' generated in Step 3. By meticulously following the overarching concept and by capitalising on the information pertaining to the resilience drivers (combinations of indicators) identified for each figure of merit and transport infrastructure mode, the development of the entire rating space was made possible.

This culminates to a tree-type approach guided by the indicator relations of Figure 4.3. A rating 'A' is assigned in each case to projects with indicator values equal or greater than those presented in the corresponding tables. A rating 'C' is assigned to cases with indicator values below the values shown in the corresponding tables. Finally, a rating 'B' (either B_{EN} or B_{EX}) is assigned for in-between values of the indicators influencing performance.

Figure 4.3 illustrates this process for the assessment of road infrastructure projects with respect to the figure of merit 'cost to completion' when FEI ≥ 0.60 and InI ≥ 0.65. The quantification process moves from left to right, starting from the exogenous indicators and gradually quantifying the influence of endogenous indicators from structural to policy. Very importantly, as the process moves from left to right, the ability of stakeholders to take action gradually increases. Figure 4.3 is also illustrative of the fact that even within a positive implementation context (FEI ≥ 0.60 and InI ≥ 0.65), adverse combinations of the endogenous indicator values (reflecting project structure) may suggest a likelihood of poor project performance (in this case for 'cost to completion'). However, it is important to remember that certain project characteristics (and therefore the corresponding indicator values) get finalised and fixed as time elapses in a project's lifetime. Consequently, a trade-off exists between the 'theoretical' flexibility that stakeholders may have based on the consideration of 'actionable' indicators, as considered by the TIRESI methodology, and the 'practical' flexibility that may be available due to the phase of implementation that the project may be in at the time of assessment. Recognising this trade-off is important to fully understand the potential of the TIRESI ratings to support decisions but also its inherent limitations.

A final point to be made is that not all indicators are included in each rating assessment, as not all indicators influence (i.e. drive resilience for) all figures of merit (outcomes). For example, in the assessment of road infrastructure projects for cost to completion when FEI ≥ 0.60 and InI ≥ 0.65, policy indicators (RAI, RRI and FSI) are not significant. This also suggests that following the project award stage when the GI, CSI and RSI indicators are finalised and fixed, there is little flexibility (or none at all) for improvements and the project exhibits the respective performance tendency ranging from A^+ to C_-.

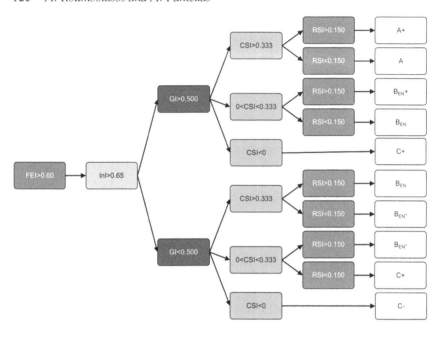

Figure 4.3 Assessment of road infrastructure projects for cost to completion for FEI ≥
0.60 and InI ≥ 0.65

Source: Authors.

The accuracy of the proposed rating system is dependent on the data that were available per mode in the BENEFIT project case dataset. The proposed system was applied to all cases, some of which were also used for the calibration of the process, based on an in-depth knowledge of their characteristics. This was of great importance when developing the rating trees for the 'actual versus forecast traffic' figure of merit, as the knowledge of which cases contained overestimated traffic forecasts was crucial.

The above approach was followed to produce the S-TIRESI ratings. Table 4.8 presents the combinations of indicators for road infrastructure and figure of merit leading to the outer limit values of the rating scale (A^+ and C_-).[2]

A point worth making from the overview of Table 4.8 is the association of poor ratings with:

- Low(er) values of the InI, stressing the need for government to invest in improving political stability, control of corruption, accountability, regulations.
- Low(er) values of the CSI, which among other factors is concentrated on the competences of actors involved in infrastructure delivery. This includes the capabilities of the contracting authority confirming the need for knowledge sharing institutions, such as the PPP units and the introduction of contractor quality assessment criteria in the contract award process.

Table 4.8 Indicator combinations and threshold values per infrastructure mode and figure-of-merit for A⁺ and C₋ ratings

Figure-of-merit	Rating	Indicator							
		FEI	InI	GI	CSI	RSI	RAI	RRI	FSI
Cost	A⁺	≥0.60	≥0.65	≥0.500	≥0.333	≥0.150			
	C₋	≥0.60	≥0.65	≥0.500	<0.000				
		≥0.60	>0.60	<0.500	<0.333	<0.150			<0.600
			≤0.65						
		≥0.50	>0.60	<0.500	<0.333	<0.150			<0.666
		<0.60	≤0.65						
		<0.50	<0.65	<0.500	<0.333	<0.150			
Time	A	≥0.60	≥0.65	≥0.500	≥0.200				
	C₋	≥0.60	≥0.65	<0.500	<0.200		≥0.500		
		≥0.60	≤0.60	<0.500	<0.200		≥0.500		≤0.600
		≥0.50	≤0.60	<0.500	<0.333	<0.150	≥0.500		
		<0.60							
		<0.50	<0.65	<0.500	<0.333	<0.150			
Traffic	A⁺	≥0.60	≥0.65	≥0.500	≥0.333	≥0.150	>0.500		
	C₋	≥0.60	>0.60	<0.600	<0.333	<0.150	≤0.500		
			≤0.65						
		≥0.60	≤0.60	<0.500	<0.333	<0.150	≤0.500		
		≥0.50	>0.60	<0.500	<0.333	<0.150	<0.500		
		<0.60	≤0.65						
		≥0.50	≤0.60	≥0.500	<0.333	<0.150	≤0.500		
		<0.60							

continued . . . .

Table 4.8 Continued

Figure-of-merit	Rating	Indicator							
		FEI	InI	GI	CSI	RSI	RAI	RRI	FSI
		≥0.50	≤0.60	<0.500	<0.333		≤0.500		
		<0.60							
		<0.50	≥0.65	<0.500	<0.000	<0.150	≤0.500		
		<0.50	<0.65	<0.500	<0.000		≤0.500		
Revenue	A^+	*Traffic Rating = A^+*							
	C_-	Traffic Rating= C	<0.500	<0.333	<0.150	<0.500	≥0.666	<0.500	
			≥0.500	<0.333	<0.150	<0.500	<0.666	<0.500	
			<0.500	<0.333		<0.500	<0.666	<0.500	

Source: Authors.

- Low(er) values of the FSI. This is not to say that publically financed projects (FSI = 1) or projects with heavy financial support by the public sector are more prone to be successful in terms of reaching their project targets. The issue is concentrated on the cost of financing, suggesting the need for low cost financing for infrastructure projects.

4.4.3 Dynamic Transport Infrastructure Resilience Indicator (D-TIRESI) quantification methodology

As mentioned at the beginning of the chapter, the D-TIRESI is assessed based on the Static (S-TIRESI). The observation of Figure 4.3 suggests that the S-TIRESI rating category of a project may change depending on the change in value of any indicator in the combination of indicators that influence the attainment of a specific figure of merit. Taking an approach that considers all such possible scenarios of changes in the relevant indicator values would be impractical and low value adding. This is because in the case of Dynamic resilience, one is mostly concerned with exogenous influences over which project stakeholders have no control.

Hence, the key scenarios considered in the assessment of the D-TIRESI reflect change in the exogenous conditions during project implementation. Of the two indicators representing the implementation context, the one presenting greater variation is the FEI, as presented in Chapter 3. Therefore, the key condition change considered corresponds to the improvement or deterioration of the financial economic context, that is the increase or decrease of the value of the FFEI. Figure 4.4 presents the variance of the FEI between 2001 and 2014 for several European countries.

Notably, variations in the FEI might bring about:

- improvement (deterioration) during the construction phase, which may influence the potential of reaching cost to completion and time to completion but also the subsequent traffic and revenue goals;
- improvement (deterioration) during the operation phase, which may influence the potential of reaching traffic and revenue goals.

An ensuing issue when estimating the D-TIRESI is the percentage of change in the FEI that needs to be considered. From the consideration of several possible approaches, this percentage has been selected to reflect the change in the FEI value that would lead to a change in the S-TIRESI rating category of the project (A, B or C), upwards or downwards. In this case, the D-TIRESI is expressed by two values, that is {X, Y}:

- a first value (X), describing the percentile decrease needed to drop to the next lower rating category, that is from A to B or from B to C; and
- rating second value (Y), describing the percentile increase needed to move up a rating category, that is from C to B or from B to A.

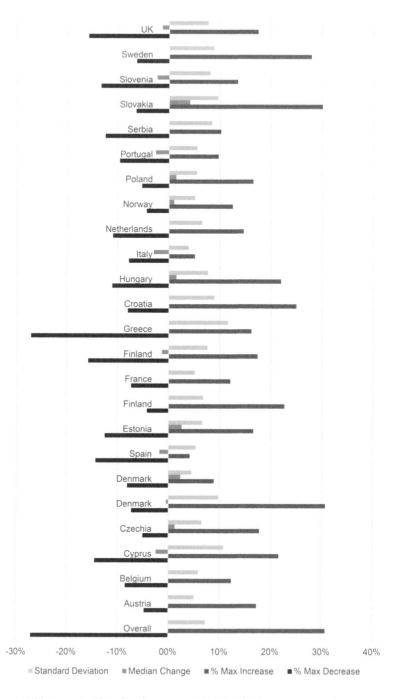

Figure 4.4 Variance in FEI values between 2001–2014 for European countries
Source: Authors.

This approach presents the advantage of providing valuable information with just two additional figures.

For example:

- Large values of percentage change indicate high stability of the S-TIRESI.
- Small values of the first value (distance from downgrading) suggest potential risk, while small values of the second value (distance from upgrading) suggest potential opportunity.
- An 'n.p.' (not possible) indication suggests that the project is not capable of withstanding fluctuations in the implementation context due to its internal structure (or because it has already reached the highest or the lowest rating category).

The concept of change in the FEI value might be extended to include changes in the InI value in cases where the InI rather than the FEI is of greater significance.

4.4.4 Overall Transport Infrastructure Resilience Indicator (O-TIRESI) quantification methodology

S-TIRESI or D-TIRESI may vary between outcomes. For all practical reasons, an overall rating (O-TIRESI) would be serving. However, as it is discussed in Chapter 6, O-TIRESI would need to express the decision maker's interests and viewpoints.

In a stakeholder-neutral approach, that is by not assuming the point of view of any project stakeholder, the presentation of the O-TIRESI is expressed as a set comprising four letter ratings, each one corresponding to one of the four figures of merit (outcomes) considered. The sequence of the four ratings will be in the form of:

$$O - TIRESI = \{Rating_{Cost}, Rating_{Time}, Rating_{Traffic}, Rating_{Revenues}\}$$

where

$Rating_{Cost}$ = Rating describing likelihood of reaching cost to completion target.

$Rating_{Time}$ = Rating describing likelihood of reaching time to completion target.

$Rating_{Traffic}$ = Rating describing likelihood of achieving forecast traffic.

$Rating_{Revenues}$ = Rating describing likelihood of achieving forecast revenue.

4.5 Operationalising Transport Infrastructure Resilience Indicator ratings

In Section 4.4, the background of the quantification methodology for the assessment of (all versions of) the TIRESI was presented. Based on the overarching principle of the quantification methodology, both exogenous and endogenous indicators are considered in the assessment of the rating of a project. More specifically, the exogenous indicators (FEI and InI) are the ones that primarily set the rating category (A, B or C) a project belongs to, with FEI being taken as the starting point of the assessment.

As noted previously, two very important threshold values for the FEI have been identified:

- FEI = 0.60
  Above this value, for all modes and all figures of merit (outcomes), the project implementation context conditions are favourable. The InI, also an important indicator of the implementation context, bears a strong correlation with the FEI and is also supported (i.e. low values are compensated) by the GI. In other words, above this threshold, the potential to reach any figure of merit (outcome) target for all modes is dependent on the internal structure of the project. In turn, the resilience of the project's internal structure depends on the values of a combination of indicators that are significant for each outcome and transport mode.

- FEI = 0.50
  When FEI ∈ [0.50, 0.60], in order to reach the respective figure of merit (outcome) targets, a project needs to be structured (designed) with endogenous resilience. This means that depending on the transport mode, the combination of other indicators (specific to each mode) define the potential resilience of the project. Their range of values, once again, depends on the mode.

  Below FEI = 0.50, the potential to reach the figure of merit (outcome) target is low, especially when combined with low values of some or all other indicators that are important for the transport mode and figure of merit under consideration. This potential may be improved depending on the structure of the project.

Considering all the information presented previously, the process that can be used to assess the rating of transport infrastructure projects can be operationalised in two (2) steps.

Step 1: The project snapshot under consideration (i.e. the project's characteristics at a specific point in its lifetime) is represented through the BENEFIT Framework indicators and their respective values.

Step 2: Ratings are assigned per figure of merit using either look-up tables, as, for example Table 4.8 in its extended form, or tree representations, as in the example of Figure 4.3.

To further facilitate the rating assessment, the TIRESI rating system is supported by a user-friendly web-based assessment tool that computes the system indicators as well as the various performance ratings based on information provided by the user. For more information on the application please visit: www.tiresias-online.com.

4.6 Conclusions

The present chapter introduces the TIRESI. The TIRESI aims to quantify 'the ability of a Transport Infrastructure project to withstand, adjust to and recover from changes within its structural elements that affect its capability to deliver specific outcomes'.

The TIRESI quantification methodology is based on the explanatory power of the BENEFIT Framework indicators and the understanding of their inter-relations as identified from the analyses presented in Chapters 2 and 3. In effect, although overarching conditions may be present, which influence the performance of many or all transport modes, there are also significant differences between the combinations of indicators and their respective values that are needed to attain specific outcomes targets per mode.

The rating system developed to support the TIRESI is detailed and transparent. It is based on considerations that facilitate the easy recognition of the likelihood of reaching pre-defined project outcomes as well as potential vulnerabilities of the project implementation system. More specifically, project rating categories have been defined as follows:

- Rating A: projects have a high likelihood of reaching a specific target outcome as they are delivered within a positive implementation context (FEI and InI). These projects demonstrate a well-structured business model (indicators IRA, CSI, RSI and GI), and supportive policy decisions (indicators RAI, RRI and FSI).
- Rating B_{EX}: projects have an average likelihood of reaching a specific target outcome as they demonstrate a well-structured business model (indicators IRA, CSI, RSI and GI), and supportive policy decisions (indicators RAI, RRI and FSI). These projects are delivered within a marginally positive implementation context (FEI and InI).
- Rating B_{EN}: projects have an average likelihood of reaching a specific target outcome as they are delivered in a positive implementation context (FEI and InI), but lack a well-structured business model (indicators IRA, CSI, RSI and GI) and/or supportive policy decisions (indicators RAI, RRI and FSI).
- Rating C: projects have a poor likelihood of reaching a specific target outcome as they are delivered in a poor implementation context (FEI and InI) and lack a well-structured business model (indicators IRA, CSI, RSI and GI) as well as supportive policy decisions (indicators RAI, RRI and FSI).

A slightly better or worse likelihood per rating is noted with a (+) or (−) notch.

Additionally, the dynamic version of the TIRESI (D-TIRESI) is assessed on top of the Static one (S-TIRESI) to determine the percentage change needed in the key implementation context indicator (FEI or InI) for S-TIRESI to move down or up a rating category. The D-TIRESI values represent the vulnerability or stability of the S-TIRESI rating with respect to the outcome target under consideration.

As highlighted from the very beginning, the BENEFIT Framework is heuristic in nature. In effect, it is built on and continuously learns from information captured from project data. In this context, all corresponding analyses and their results bear the limitations of the original project case dataset in terms of its accuracy, level of detail, and breadth of transport modes covered. Effectively, the BENEFIT project case database has been built by collecting information on projects, which was readily available in the public domain.

Notably, while the presented indicators and their quantification methodologies may be continuously improved over time through their application on a wider and more detailed sample of projects, the current downside ramifications are that:

- The data the indicators have been built upon lack detail that in many cases hampered the ability to conduct more advanced and detailed analyses. This limitation was addressed by conducting multiple parallel analyses through which different aspects and conditions could be identified.
- In its present state of development, a reliable rating system cannot be provided for the following modes of transport:

  – rail infrastructure, as the information made available was not sufficiently conclusive, even though the analysis of the literature suggests that the performance of rail infrastructure could be captured by the same indicators identified for urban transit projects but with the additional consideration of the FEI;
  – airport infrastructure, with respect to attaining actual versus forecast traffic and revenue outcome targets. For this mode, the FEI was also unable to fully represent the implementation environment for airport operation;
  – port infrastructure, with respect to all project outcomes due to the characteristics of this mode in combination with the impact of strategic behaviour exhibited by port operators.

On the positive side, the fact that detailed data were not required to develop and quantify the various indicators and estimate the TIRESI ratings, makes the following actions possible:

- conducting ex-ante scenario analyses (as detailed data are not available at that stage), and combining them with other existing methods used in the pre-award phase;
- using the TIRESI as a monitoring tool during project implementation;
- introducing improvements in the project structure and identifying the potential effects of changes in the funding and financing schemes;
- assessing the effectiveness of improvement measures during project implementation;
- enabling project assessment by non-technically experienced stakeholders.

Examples of the above possible actions are provided in Chapter 5 through the consideration of nine project cases ranging across different modes.

The assessment of the above cases will also confirm the competence of the actors involved in project delivery, also identified through the rating tables concerning the importance of mature institutions, as well as the need for low-cost financing.

Finally, in practical terms, TIRESI ratings reflect likelihoods and not an absolute certainty. Consequently, project managers should consider a poor rating as a warning sign reflecting the need for close project monitoring and effective risk management. More importantly, the TIRESI ratings are meant to be used as a guiding tool to be employed by all stakeholders throughout the project lifetime. Ratings should also be used in correlation with the identified, in each case, resilience drivers to support decisions on actionable project elements and factors. Obviously, certain project characteristics (and therefore the corresponding indicator values) are difficult or undesirable to change as time elapses in a project's lifetime. Rating tables may then be useful in assessing trade-offs. Such stakeholder-specific considerations are further discussed in Chapter 6.

4.7 Notes

1   A "figure-of-merit" is a time-dependent measure of the level of delivery of the system with respect to its intended service (Henry and Ramirez-Marquez, 2012)
2   For all transport modes and figures of merit, please see www.teresias-online.com

4.8 References

Bruneau, M., Chang S., Eguchi R., Lee G., O'Rourke T., Reinhorn, A., Shinozuka, M., Tierney, K., Wallace, W., and von Winterfeldt, D. (2003), A Framework to Quantitatively Assess and Enhance Seismic Resilience of Communities. *Earthquake Spectra*, 19(4), pp. 733–752.

Filippini, R., and Silva, A. (2014). A Modeling Framework for the Resilience Analysis of Networked Systems-of-Systems Based on Functional Dependencies. *Reliability Engineering and System Safety*, 125, pp. 82–91.

Fisher, R.W., Bassett G.W., Buchring, W.A., Collins, M.J., Dickinson, D.C., and Eaton, L.K. (2010). Constructing a Resilience Index for the Enhanced Critical Infrastructure Protection Program, Report ANL/DIS-10-9, Argonne National Laboratory.

Francis, R., and Bekera, B. (2014). A Metric and Frameworks for Resilience Analysis of Engineered and Infrastructure Systems. *Reliability Engineering and System Safety*, 121, pp. 90–103.

Henry, D., and Ramirez-Marquez, J.E. (2012). Generic Metrics and Quantitative Approaches for System Resilience as a Function of Time. *Reliability Engineering and System Safety*, 99, pp. 114–122.

MERRIAM-Webster (n.d.), Origin and definition of resilient. www.merriam-webster.com/dictionary/resilient [Accessed 9 May 2016].

Oxford Dictionaries (n.d.), Definition of resilience. www.oxforddictionaries.com/definition/english/resilience (Accessed 9 May 2016).

Ragin, C. C. (2000). *Fuzzy-Set Social Sciences*. Chicago, IL: University of Chicago Press.

Ragin, C.C. (2008). *Redesigning Social Inquiry – Fuzzy-Set and Beyond.* Chicago and London: The University of Chicago Press.

Rose, A., and Krausmann, E. (2013). An Economic Framework for the Development of a Resilience Index for Business Recovery. *International Journal of Disaster Risk Reduction*, 5, pp. 73–83.

Rose, A. (2015). Measuring Economic Resilience: Recent Advances and Future Priorities, Conference on 'Effective Corporate Leadership and Governance Practices in Catastrophe Risk Management'. Philadelphia, PA: Wharton School, University of Pennsylvania.

Roumboutsos, A., Pantelias, A., Sfakianakis, E., Edkins, A., Karousos, I., Konstantinopoulos, E., Leviäkangas, P., and Moraiti, P. (2016). Deliverable D3.2 – The Decision Matching Framework Policy Guiding Tool, Project Rating Methodology and Methodological Framework to Increase Business Model Creditworthiness, BENEFIT (Business Models for Enhancing Funding and Enabling Financing for Infrastructure in Transport) Horizon 2020 Project. Grant Agreement No. 635973. www.benefit4transport.eu/index.php/reports.

5 Analysing scenarios of transport infrastructure funding and financing

Hans Voordijk

5.1 Introduction

In this chapter, the use of the TIRESI rating system is demonstrated in nine European transport infrastructure projects.

The scope is to demonstrate the BENEFIT rating system's potential to:

- predict project performance;
- introduce improvements in the project structure and identify the potential effects of these changes;
- assess the effectiveness of measures during project implementation (e.g. during renegotiations);
- test alternative funding schemes;
- enable the assessment of alternative financing schemes;
- address both public and private (co) financed projects.

In addition, the limitations and extended abilities of the TIRESI rating system are demonstrated.

Each case study consists of two parts. The first part (project delivery) describes the actual project delivery in terms of implementation context, contractual governance, business model efficiency, funding and financing as well as project outcomes. The second part (alternative scenarios of project delivery) focuses on actions or measures that could have improved the outcomes of the project. In this context, hypothetical scenarios of project development are tested. As a reminder, when considering the description of cases through indicators, projects are represented by a series of snapshots that show the 'status' of their delivery and/or implementation at particular points in time of their life cycle. The 'status' is captured through the indicators values of the BENEFIT Framework at these particular points in time.

A brief outline of information included in each project description follows.

5.1.1 Project delivery

In this part, the focus is on the ex-post analysis of project delivery. More specifically, the project's development over time is presented. Through this description, the respective indicators and their values are introduced. Each project description follows upon the BENEFIT Framework.

Context

The focus is on the 'raison d'être' and the implementation context of the project. For each project, the context is provided in terms of: a project timeline, the national productivity and prevailing macro-economic conditions (FEI) and government stability (InI). The time (year) of each snapshot is also depicted for convenience and greater understanding.

Contractual governance

The focus is on the governance of the project. Attention is devoted to the tendering and procurement process, type of contract and how and by whom construction and operation is delivered. Risk allocation between the public and private parties is discussed for the following risks: design and construction, maintenance, exploitation, commercial revenue, financial, regulatory and force majeure. In the projects studied, certain risks are assumed by public parties, others by private parties, while some are shared. Based on this analysis, values are assigned to the GI throughout the project's lifetime.

Business model efficiency

The focus is on the project's business model. Three key aspects are discussed here. First, attention is devoted to project integration, that is the position of the project in the existing transport network and how this is supportive or not of its exclusivity. Second, the competences of the contracting authority and the private sector actor(s) involved in the project are discussed. Based on this analysis and, in combination with existing evidence on the project's risk allocation, the values of the CSI and the RSI are presented for the respective project snapshots.

Funding and financing

The focus is on the policy tool indicators. Funding deals with the presentation of project revenues and the project's remuneration scheme, referring to the RRI and the RAI, respectively. Financing focuses on the description of the sources of capital committed to the project and informs the computation of the FSI.

Outcomes

The focus is on project outcomes. The observed outcomes of each project are stated referring to cost and time overruns (or underruns) to construction completion, forecast versus actual traffic, and forecast versus realised revenues.

Ultimately, for every project studied, the ex-post analysis of its delivery is summarised in one comprehensive table that contains the aforementioned indicator values as well as the project outcomes (in terms of cost, time, traffic and revenues), and is compared to the TIRESI rating for each outcome and snapshot, as this results from the respective indicator values.

5.1.2 Alternative scenarios of project delivery

In the second part, the focus is on the analysis of the projects studied in terms of identifying actions or measures that could have improved their outcomes. Hypothetical scenarios of project development are also tested. Through this analysis, the use of the TIRESI rating system is demonstrated as an ex-ante project assessment tool. This rating may assist in predicting project performance and adopting measures during the project life in order to minimise adverse effects from the change of external or internal project conditions.

More specifically, in this part, the factors driving performance are identified and reflected on combinations of indicators and their respective values that would need to be changed accordingly. In this context, it is possible to identify:

* Which of the project outcomes can be improved and which of the elements increase the likelihood of meeting these project outcomes the most?
* Which are the most realistic opportunities to achieve the desired outcomes?
* Which actions are suitable and necessary to implement in the project under consideration and their specific description?

In addition, 'what if' scenarios can also be tested. Finally, conclusions are drawn for each case study based on the limitations/possibilities for improving the project.

5.1.3 Case studies

Table 5.1 Overview of case studies

#	Case name	Country	Mode concerned	Financed publicly (construction)
1	Combiplan Nijverdal	The Netherlands	Rail/Road	✓
2	Motorway Horgos–Novi Sad	Serbia	Road	✓
3	The Blanka Tunnel Complex	Czech Republic	Tunnel	✓
4	Reims Tramway	France	Urban: tram	
5	Brabo 1	Belgium	Urban: tram	
6	E39 Klett-Bardshaug Road	Norway	Road	
7	A5 Maribor Pince Motorway	Slovenia	Road	✓
8	Athens Ring Road	Greece	Road	
9	Port of Agaete	Spain	Port	✓

Source: Authors.

The nine project case studies that have been studied in this chapter pertain to the following transportation modes: road (four), road/rail (one), urban transport (two), tunnel (one) and also a seaport as a special application. Four projects are publicly financed, while the remaining five have been procured under PPP arrangements.

5.2 Combiplan Nijverdal, The Netherlands

Hans Voordijk and Ibsen Chivata Cardenas

5.2.1 Project delivery

5.2.1.1 Context

The Combiplan project, financed through public sources, included the construction of a new rail and motorway across the centre of the Nijverdal city over a distance of 6 kilometres, as part of a transportation link connecting the cities of Zwolle, Nijverdal and Almelo in the Netherlands. The infrastructure built is connected to the central train station. A 500 metres length tunnel that accommodates two train tracks and 2 by 2 lanes is part of the project. The tunnel alignment goes along the old railway path. The width of the tunnel structure ranges from approximately 26 to 36 metres. The tunnel depth varies along the route between 6 and 10 metres. Local infrastructure was relocated for the new project. Two slow traffic depressed crossings, two road bridges, two railway bridges, a railway viaduct and the necessary noise barriers were also built.

At the time the project was planned and constructed, the socio-economic conditions were favourable. Following project award in 2006, the GDP growth rate increased. Comparing the years of planning (between 1990 and 1998) with those during construction (2005–2012), there was an incremental variation beyond expected values of the per capita income level of the region. Likewise, the unemployment rate dropped after 2006 (contract award year).

Some other characteristics of the market geography in the region show that the population density at the time of the project award was higher than the one during the project planning stage. Similar trends were observed for the level of industrialization and economic activities at the time of data collection of the project with respect to industrialization at project award. No new specific production activities started during or after the completion of the project and none was foreseen in the planning stage of the project.

Figure 5.1 illustrates the project's development (project timeline, IRA, FEI and InI). The figure also depicts the project's four snapshots considered in the calculation of the indicator sets.

5.2.1.2 Contractual governance

For the project, the original initiative was to undertake it using a PPP scheme. Eventually, it was decided to award a traditional design and construction contract, which was common practice. At the time, the national government did not stimulate much the use of PPP contracts.

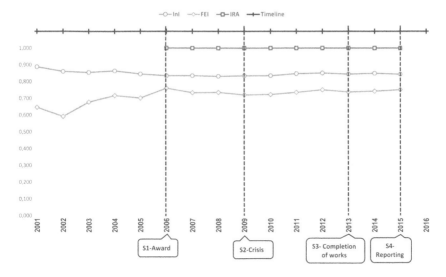

Figure 5.1 Combiplan Nijverdal project timeline
Source: Authors' compilation.

The contracting authority awarded a design and construction contract. However, after some contract modifications, maintenance services were included. The maintenance service period was agreed to be 10 years for both the road and the accompanying information transport system. The rail infrastructure attached to the project is to be maintained by ProRail, a Government agency, through outsourcing. The Nederlandse Spoorwegen (NS) is responsible for the rail operation, and certain operation activities are also outsourced. NS is a state-owned company in charge of the operation of Dutch railways. Rijkswaterstaat (RWS), an agency of the Ministry of Infrastructure and Environment in the Netherlands, takes care of the maintenance and operation of the roads attached to the project. In other words, following the joint design and construction of a rail and road section (Combiplan project), operation was resumed by the responsible public sector entities, respectively.

For this project, the procurement process consisted of a two-stage procedure in which there was an open call first and, following this, a competitive dialogue procurement procedure. Five bidders participated in the tendering process, which lasted 12 months.

Unit costs is the payment basis for the contract. The contract duration was agreed to be 68 months.

The respective allocation of risk is presented in Table 5.2. It should be noted that the contracting authority awarded a design and construction contract. Since ProRail and NS, for rail infrastructure and operation, and RWS, for road operation, are responsible for maintenance and operation and are public sector entities, it can be said that revenue, financial, exploitation, regulatory and force majeure risks are mostly borne by the public party.

Table 5.2 Combiplan Nijverdal – risk allocation

Risks	Totally private		⟺			Totally public
Design			✓			
Construction		✓				
Maintenance			✓			
Exploitation					✓	
Commercial/revenue					✓	
Financial					✓	
Regulatory				✓		
Force majeure				✓		

Source: Authors.

Contractual governance in this project exhibits a low degree of implementation considering the governance elements proposed by the BENEFIT Framework. The granting of incentives for performance, the collective investments estimation and the enforcing of termination clauses were options not fully set in place for the project. This is also evident in the values of GI, which were 0.375 for 2006–2013 and 0.500 for 2015.

Enhancing governance is very constrained given the exceptional characteristics of the project under analysis. There are three players in the operation phase, two organisations maintaining road and rail civil works and another that operates the rolling stock. All organisations are legally public bodies. This leads to the situation in which there is not any cooperative approach to project delivery, which is the core tested assumption for project governance as analysed in the BENEFIT project. Furthermore, the allocation of risk is fully concentrated on the public party without the immediate possibility of being transferred elsewhere in the current state of the project (operation).

5.2.1.3 Business model efficiency

PROJECT INTEGRATION

The project enables freight and passenger mobility between the city of Zwolle and the Twente region. Road and railways benefit Nijverdal citizens directly, as well as residents and visitors from the Zwolle, Nijverdal and Almelo municipalities. Since the Combiplan project is highly connected with the Dutch transport network, it involves other regional and national users.

COMPETENCE OF THE CONTRACTING AUTHORITY

RWS is the Government agency and national authority attached to the Ministry of Infrastructure and Environment. RWS invests about 3 bn EUR a year. ProRail is the government agency, which carries out the maintenance of the national railway network infrastructure, provides rail capacity and controls traffic. In 2013, ProRail invested 1.2 billion EUR in rail and stations infrastructure. RWS in conjunction with ProRail are the client and public party, respectively, involved in this project. These two contracting parties have not shown major problems administrating the respective assets that provide an indication of their competence.

COMPETENCE OF THE PRIVATE SECTOR ACTOR(S)

The contractor responsible for the design, construction as well as road maintenance for the first ten years after the full completion of works is a consortium consisting of Van Hattum and Blankevoort, Hegeman Beton-en Industriebouw, Koninklijke Wegenbouw Stevin BV (KWS) and VolkerInfra Systems (Vialis). The design of works was executed by VolkerInfra Design. Royal Haskoning DHV acted as advisor to the consortium and to ProRail. Hans van Heeswijk Architects delivered the architectural design. The contracting companies involved in this project have a good reputation in the Netherlands and show revenues above the average of companies in Europe in the last years.

Despite the satisfactory background of the contractors and client, eventually, the project failed to be delivered on time and experienced cost overruns. A number of situations affected the project. For instance, scope changes occurred with some additional works promoted by the municipality and the province. The overall increase of the contract scope from the start to the end of the works amounts to about 30 per cent. Cost overrun was due to increased scope (30 per cent), new design requirements (40 per cent) and changes due to new tunnel standards (30 per cent).

The resulting delay of the opening was due to connection problems in the network between the traffic control centre and the tunnel. The delay was approximately six to nine months.

Using the elements studied in the BENEFIT project, two direct causes can be associated with the problems that occurred in the project. It might be asserted that the introduction of technology innovations was not optimal and, respectively, the contractor did not have sufficient past experience with the innovative information technologies being implemented. Next, the contracting authority could have also failed to optimally manage stakeholders, since the project underwent changes to the scope due to possibly the untimely or inappropriate consideration of some of the stakeholders' expectations.

Based on the above, and in combination with the risk allocation structure, as presented in the previous section:

- the CSI was 0.343 at the award (2006), decreased to 0.247 during crisis (2009) and at completion of works (2013), and then increased to 0.278 in 2015;
- the RSI remained 0.214 for all snapshots.

5.2.1.4 Funding and financing

The construction of the project was financed through public sources. The national government contribution was 272 M EUR (2009). In addition to this budget, there are regional contributions as follows:

- 5.2 M EUR by Nijverdal (village of the municipality Hellendoorn);
- 16 M EUR by Zwolle municipality;
- 14.6 M EUR by the Overijssel province;
- 30 M EUR from the Region of Twente (association of certain municipalities in the area of influence).

Fares, subsidies and public funds (roads) are the main streams of income during operation, while the public budget financed the construction phase. In the operation phase, public agencies or state-owned companies are responsible for collecting fares and bearing the remuneration risk. In 2015, in the Netherlands, 88 per cent of the national income of the rail operator was fares collected, while approximately 12 per cent came from the exploitation of rail stations. Resources for road maintenance come from taxation.

The funding scheme in this project appears to be considerable robust, although it could be improved by increasing the control of demand by means of generating an exclusive use of the infrastructure, being this an option considered by the framework proposed by the BENEFIT project. Likewise, possibly more can be done to increase the proportion of non-transport remuneration sources and, by doing so, the robustness of the remuneration scheme.

Based on the above:

- the RAI was 0.833 throughout the project lifetime;
- the RRI also remained the same (0.833) throughout the project lifetime.

FINANCING SCHEME

No debts were required for this project and this explains the robustness of its financing scheme, as only public sector funds and government subsidies were employed.

Based on the above, the FSI was 1000 throughout the project lifetime.

5.2.1.5 Outcomes

COST AND TIME TO CONSTRUCTION COMPLETION

The project had a relatively long planning period (1990–2006). During this period, project scope and budget were adjusted. Changes were made following project award.

The opening of the road tunnel was delayed for approximately six to nine months due to the need to implement the new national tunnel standard and connection problems in the network. The impact is about 10 per cent of the total forecasted lead time.

Three situations were handled as 'claims'. These refer to the delay and acceleration costs and changed laws and regulations (e.g. red diesel could not be used on the building site).

ACTUAL VERSUS FORECAST TRAFFIC

Despite the above setbacks, the road tunnel is expected to reduce travel time, improve traffic flow and road safety on the N35 and at the same time the quality of life in the centre of Nijverdal, while traffic forecast targets are met.

PROJECT RATING OVER TIME

Table 5.3 presents the results of the analysis for the Combiplan project, based on the indicators proposed in BENEFIT project, the project outcomes and the 'TIRESI' ratings over time.

Ratings reflect the less than favourable governance of the project and properly predict the cost and, especially, time overrun. The project was implemented and, then, operated under favourable implementation context conditions (FEI and InI) allowing for favourable traffic and revenue ratings.

5.2.2 *Alternative scenarios of project delivery*

According to the BENEFIT Framework, the analysis of the Combiplan project showed some potential of improvement of the GI and CSI, as follows:

1   the granting of incentives for performance, which are expected to (i) discourage opportunistic behaviour on the part of the contractor, and (ii) may motivate the contractor to minimize project duration;
2   the collective investments estimation would contribute to (i) reducing uncertainty in the project; (ii) facilitating contractors' contributions in the design stage that may establish a trust-based cooperative relationship; (iii) providing complete plans and specifications that may decrease the number of disagreements and disputes; (iv) maximising outcomes, since the owner benefits from a range of design scenarios, sensible risk management (risk awareness) and appropriate contract development;
3   enforcing contract termination without cause clauses would grant unilateral control to the client, reducing the likelihood that a contractor can engage in opportunistic rent seeking. The possibility of unilateral termination by the client correspondingly lowers the likelihood of underinvestment by the contractor, facilitating smooth adaptation to unfolding contingencies;

4   the introduction of technology innovations should have been considered, if and only if, the contractor had previously shown successful innovations applications;

5   stakeholders management by the public authority should have been considered, if and only if, the authority had managed stakeholders in equivalent projects;

6   the optimisation of the control of demand by generating an exclusive use of the infrastructure;

7   likewise, possibly more could be done to increase the proportion of non-transport remuneration sources and by doing so the robustness of the remuneration scheme.

All the above seven actions were open options for the project at some point, yet some inherent constraints could have hampered their implementation. For instance, some constraints were the little consideration of the potential of cooperation-based approaches to project delivery (e.g. PPP), and the reliance on traditional procurement at the moment the project was procured. These possibilities of improved performance meet further constraints during the operation of the project due to the fact that project operation is currently done by public agencies or state-owned companies usually advocating traditional procurement approaches.

Following on the above, one hypothetical scenario was tested: improvement of contractual governance. The scenario incorporated the options (1) to (3), described above, for the award year, as well as suggestions (4) and (5) concerning the successful application of innovation and better stakeholder management. As expected and given the very positive implementation context and the supportive financing and funding scheme, the project ratings improved significantly featuring A$^+$ across all outcomes apart from time to completion that was rated A.

5.2.3 Concluding remarks

The Combiplan project featured the combined design and construction of road and rail infrastructure with operation, following project completion, by the respective public authorities responsible for rail and road operation and maintenance in the Netherlands. The project was implemented in the favourable implementations context conditions characterising the country, while the infrastructure (road and rail) was built to serve a growing population.

The key impediment of the project was in contractual governance leading to both cost and time to construction completion overrun.

The TIRESI ratings correctly illustrated the project's actual development.

The tested hypothetical scenario concerned improvements in contractual governance, which were identified in the case analysis. The resulting project ratings of this scenario reflected the expected improvement.

Table 5.3 Combiplan Nijverdal – ex-post analysis and alternative scenarios

Snapshot	Indicator value									Cost		Time		Traffic		Revenue	
	FEI	InI	GI	CSI	RSI	RAI	RRI	IRA	FSI	Out-come	TIR-ESI	Out-come	TIR-ESI	Out-come	TIR-ESI	Out-come	TIR-ESI
Snapshot 1 – Award (2006)	0.760	0.835	0.375	0.343	0.214	0.833	0.833	1.000	1.000	N.A.	B_{EN}	N.A.	C^+	N.A.	B_{EN}	N.A.	A_-
Snapshot 2 – Crisis (2009)	0.721	0.835	0.375	0.247	0.214	0.833	0.833	1.000	1.000	N.A.	B_{EN-}	N.A.	C^+	N.A.	B_{EN-}	N.A.	B_{EX}
Snapshot 3 – Completion of works (2013)	0.737	0.845	0.375	0.247	0.214	0.833	0.833	1.000	1.000	–1	B_{EN-}	–1	C^+	–1	B_{EN-}	0	B_{EX}
Snapshot 4 – Reporting (2015)	0.751	0.844	0.500	0.278	0.214	0.833	0.833	1.000	1.000	–1	N.R.	–1	N.R.	–1	B_{EN}^+	0	A_-
Alternative scenarios																	
Scenario 1 – Award with improved GI/CSI	0.760	0.835	0.688	0.361	0.214	0.833	0.833	1.000	1.000	N.A.	A^+	N.A.	A	N.A.	A^+	N.A.	A^+

Legend: N.A. = not available; N.R. = not relevant
Source: Authors.

5.3 Motorway Horgos – Novi Sad (second phase), Serbia

Miljan Mikić, Jelena Cirilovic, Nevena Vajdic, Nenad Ivanisevic and Goran Mladenović

5.3.1 Project delivery

5.3.1.1 Context

This case study presents the construction of 107 kilometres of the second (left) carriageway of the second phase of construction of the E-75 (M-22) motorway from the Hungarian border (Horgos) to the city of Novi Sad, the capital of the Serbian province Vojvodina (Figure 5.2). The Horgos – Novi Sad motorway consists of four lanes (two in each direction), two emergency lanes and a central reserve, as well as 13 pairs of bridges, nine interchanges, two main toll gates (one already completed), three large fills, 44 pipes, frame and box culverts, four rest areas (two already constructed) and several multipurpose lay-bys with gas stations and motels.

The motorway Horgos – Novi Sad was included in the scope of works in the planned Horgos – Pozega Toll Motorway concession, which was cancelled in 2008 due to difficulties in reaching financial close. The scope of work was, thereafter, procured in sections and financed by the public budget. Two sections have already been completed in 2009 (from 28 to 38 kilometres and from 98 to 108 kilometres) on a design–bid–build contractual basis. In 2011, construction of the remaining 86.88 kilometre carriageway was completed on the same basis and the section from Horgos to Novi Sad was opened to traffic (Figure 5.2). The present description concerns sections 0–38 kilometres and 38–98 kilometres, which were tendered under one contract, hereafter Horgos – Novi Sad project.

Figure 5.3 illustrates the project's development (project timeline, IRA, FEI, and InI). The graph also depicts the project's three snapshots.

Figure 5.2 Motorway Horgos – Novi Sad route
Source: Authors' compilation.

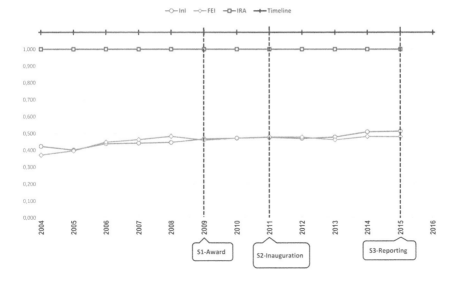

Figure 5.3 Motorway Horgos – Novi Sad (2nd phase) project timeline
Source: Authors' compilation.

5.3.1.2 Contractual governance

The former Ministry of Infrastructure (now Ministry of Construction, Transport and Infrastructure) was the contracting authority. Project coordinator (project implementation unit) was the company, Koridori Srbije Ltd., founded in 2009 by the Government of the Republic of Serbia, with the goal to improve the efficiency of construction on the new road infrastructure projects in the Republic of Serbia.

The contractual basis for project delivery was a design–bid–build contract with unit price payment. No standard form of the contract agreement was used. The Roads Authority of Serbia takes over the responsibility for operation and maintenance following guarantee period. This was delayed due to incomplete works in one rest area.

The open call for tender was launched in February 2010. There were six bidders that participated and three bidders who submitted their offers. However, the results were annulled. The procedure was then changed to negotiations with the Serbian consortium, which offered the lowest price in the previously annulled tender.

RISK ALLOCATION

All sources of risk except for Construction and Force majeure were allocated to the public side. This was due to the fact that the project was financed from a public budget and delivered as a design–bid–build contract, with the public side responsible for design, maintenance and exploitation as well as commercial and

Table 5.4 Horgos-Novi Sad Motorway – risk allocation

Risks	Totally private	⟺		Totally public
Design				✓
Construction		✓		
Maintenance				✓
Exploitation				✓
Commercial/revenue				✓
Financial				✓
Regulatory				✓
Force majeure			✓	

Source: Authors.

regulatory conditions during construction and operation. Design–bid–build is a contract model used for construction contracts on almost all sections of the motorway network designed and built in Serbia in the last 15 years. If there were (as was commonly the case) external loans used for financing the project under this contract model, the obligation of the Serbian public side was to conclude land acquisition (through expropriation) and to provide the final design for the project. Major sources of risk that caused time and cost overrun on almost all motorway projects in Serbia were related to lack of properly and timely conducted expropriation on certain parts along the route, as well as poor design. The risks with the highest impact on time and cost were actual hydrogeological and geotechnical conditions being significantly different from the designed ones.

Based on the above, the GI constantly has a very low value of 0.19 throughout the project construction period, but increased to 0.500 following the start of operation.

The risk allocation is as depicted in Table 5.4.

5.3.1.3 Business model efficiency

PROJECT INTEGRATION

The scope of the second phase of the Motorway Horgos – Novi Sad was a completion of the motorway that would offer:

- connection of the Hungarian–Serbian border (Horgos) to the city of Novi Sad, the capital of Serbian province Vojvodina;
- the segment of European route E-75 (former Pan European Transport Corridor X) which connects Finland, Poland, Czech Republic, Slovakia, Hungary, Serbia, Macedonia and Greece. Nowadays, it is marked as a comprehensive TEN-T (where TEN is Trans-European Network) corridor;
- reduction of travel time trough increase of operating speeds;
- reduction of vehicle operating costs;
- reduction of number of traffic accidents caused by inadequate cross-section of semi-highway.

COMPETENCE OF THE CONTRACTING AUTHORITY

As stated previously, the contracting authority was the Ministry of Infrastructure, while the Project Coordinator was the public company Koridori Srbije Ltd.

The contractor selection process according to the public procurement law in Serbia is strictly based on the criterion of the lowest bid price. This criterion in combination with high competition in the construction market (significantly affected by the economic crisis from 2008 onwards) regularly results in the under-estimation of costs and time.

However, the process of rebuilding the Serbian motorway network has been intensified since 2007. Consequently, the capacity of the contracting authority to plan, contract, manage and monitor capital construction projects and processes has been increased since then, but can be assessed as limited for the period 2009–2011, when the Motorway Horgos – Novi Sad was built.

COMPETENCE OF THE PRIVATE SECTOR ACTOR(S)

A consortium consisting of the following local contractors was awarded the project: PZP Beograd, AD Putevi Uzice, GP Planum, Borovica transport. Two of the companies in the consortium (AD Putevi Uzice and GP Planum) were internationally renowned Serbian companies in the motorway construction sector, while the other two were local companies, previously dealing mostly with road maintenance works.

Due to a specific contract condition, explained further in this case study, one of the consortium members, PZP Beograd, went into bankruptcy 12 months into project construction.

Based on the above information and in combination with the risk allocation structure as presented in the previous section (see Table 5.5):

- the CSI was 0.092 at the award time (2009), 0.123 at the start of operation (2011) and 0.000 at the reporting time (2015);
- the RSI was 0.228 in all project phases.

5.3.1.4 Funding and financing

Unlike the other sections of E-75 (M-22) motorway, for which loans were provided from International Financing Institutions, the section Horgos – Novi Sad was financed entirely through the public budget.

Apart from minor revenues connected to land use and advertisements, the remuneration and revenue scheme is based on tolls. However, since project inauguration the toll collection was not very efficient since an open toll section already existed. Only recently has the toll collection been converted to a closed system including the remaining part of motorway leading to Belgrade.

The RAI had a value of 0.367 and then improved to 0.667 when the toll regime changed to a closed system. Likewise, RRI had a value of 0.563 that improved to 0.607 and finally 0.833, reflecting the change in the tolling risk profile and cost coverage. The FSI also had a constant value of 1000 across all project phases (see Table 5.5).

5.3.1.5 Outcomes

According to the data available from the project and the data obtained from an interview with the investor personnel, the reliability, availability and safety have been improved in line with project expectations.

As for the outcomes considered, the project experienced cost and time overruns, the actual traffic is lower than forecasted, while revenues are in line with forecasts.

COST TO CONSTRUCTION COMPLETION

The actual construction costs on the project were 25.9 per cent higher than the original budget. For one of the sections (from 00 to 28 kilometres), there is still an unresolved lawsuit about the cost adjustment. The value requested by the contractor in the lawsuit is 13 M EUR, which comes on top of the stated cost overrun (25.9 per cent). Despite under-performance in terms of cost, it has to be noted that investor's personnel considered the actual final construction costs of 1.72 M EUR/kilometre as a very good achievement on the project.

Nevertheless, the specific payment method envisaged in the contract over-stretched the contractors' self-financing capability and lead one of the consortium members to bankruptcy.

TIME TO CONSTRUCTION COMPLETION

There was one year of delay in the construction phase of the project. The main reason for the delay was that the contractual time for completing the construction works on sections from 00 to 28 kilometres and from 38 to 98 kilometres was defined under political pressure as nine months. Technical estimation of the time needed for completion of those two sections was 18 months. Execution of works actually lasted 21 months.

An additional very important reason for the delay was the fact that one of consortium members (PZP Beograd) went into bankruptcy 12 months after the commencing of works.

ACTUAL VERSUS FORECAST TRAFFIC

Traffic was overall assessed as in line with projections, since the actual traffic was around 20 per cent lower than estimated.

Figure 5.4 presents the forecast and actual average annual daily traffic (AADT) on the motorway since 2005. Despite the strong impact that the global financial crisis had on the country economy (that is also reflected through the decreasing

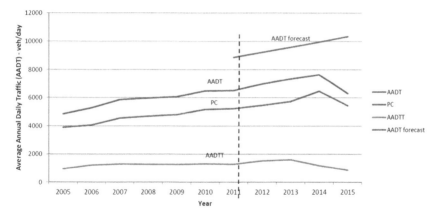

Figure 5.4 Traffic on Motorway Horgos – Novi Sad from 2005 to 2015
Source: Authors' compilation.

trend of the FEI indicator in Figure 5.3), the traffic had an increasing trend throughout the whole period, although around 20 per cent lower than forecast AADT until 2015. Final figures have not been released yet, but the lower figures may have been influenced by the refugee crisis that had an impact on the traffic on the Serbia – Hungary border crossing.

PROJECT RATING OVER TIME

Table 5.5 presents the results of the ex-post analysis for the Motorway Horgos – Novi Sad project, based on the indicator values, the project outcomes and the 'TIRESI' ratings over time. Notably, project ratings for all outcomes coincide with observed project performance.

5.3.2 *Alternative scenarios of project delivery*

In the case of Motorway Horgos – Novi Sad, GI has constantly a very low value of 0.19 and is surely a negative sign for the project potential to achieve cost, time and traffic targets. In addition, CSI values have remained very low: 0.092 at the award time, 0.123 at the start of operation, reaching 0.000 at the reporting time. Both GI and CSI contribute significantly to the low rating of the project.

Policy indicators changed over time reflecting the changes introduced to the tolling regime.

Based on the above, the major factors influencing performance originate from the project planning and procurement phases. This, as stated, reflects very well not only the present case study, but also the general Serbian context, characterised by low institutional maturity and low capability of the monitoring authority.

Realistic options for improvement concern indicators GI and CSI (GI at 0.19 and CSI at 0 to 0.123) for cost and time expectations.

To improve governance, it would be necessary to introduce incentives for performance to the contractor, which would enable a more 'partnering' relationship between a client and a contractor, along with the traditional measure of penalty for not achieving a target deadline. Involving more bidders and integrating design and construction as conditional measures are also recommended, but could be applied only after certain preconditions are met (e.g. a change in the public procurement legislation to allow for selection criteria other than lowest price).

Furthermore, since neither a design-build contract model nor any of the PPP models have been applied in the motorway construction in Serbia so far, public authorities' capability to implement and monitor such contracts is currently assessed only as limited.

Therefore, the involvement of more bidders and the integration of design and construction are suggested measures to be applied for improving Governance, when the following preconditions are met: (i) introducing several criteria into the practice of bidders evaluation in an open-call procedure (in contrast to the current practice of the lowest price only) including technical capacity and evidence of previous performance (precondition to be met for a successful implementation of the first suggested measure); (ii) building up the capacity of a public company for managing design and build contracts (precondition for the second measure).

To improve the cost saving potential, the requirement of adequate references may be considered, for the contractor as a company, and for its key personnel that would prove their performance in previous similar projects and their capability to deliver the current project. In addition, this requirement should also cover the Contractor's capability to adequately monitor construction. In order to efficiently manage a project, the public partner would need to have past experience on similar projects, or to engage a consultant with similar references.

It can be seen that in the award stage, if measures for improving GI and CSI are applied, then GI would take a value of 0.500 (compared to 0.188 in the base scenario), CSI would become 0.271 (compared to 0.092 in the base scenario) and the project rating regarding cost and time will improve from C (base scenario) to B_{EX-}. Project ratings regarding other performance criteria (traffic and revenue) are also slightly improved compared to the base scenario.

In the inauguration stage, if appropriate measures are applied CSI becomes 0.271 (compared to 0.123 in the base scenario), and the GI value improves to 0.500, which is reflected on the improved ratings for cost and time to completion.

At the reporting time (2015), the application of necessary measures enables an improvement in the value of the CSI, which becomes 0.444 (compared to 0.000 in the base scenario), yielding increased project rating results for traffic and revenue outcomes.

To summarise, in the ex-ante scenario analysis, potential measures for improving the governance and cost saving potential were identified and analysed. If measures evaluated as the most effective would be applied, the GI would have a higher value in the award stage and the CSI would increase in all project stages.

Table 5.5 Horgos-Novi Sad Motorway – ex-post analysis and alternative scenarios

Snapshot	Indicator									Cost		Time		Traffic		Revenue	
	FEI	InI	GI	CSI	RSI	RAI	RRI	IRA	FSI	Outcome	TIR-ESI	Outcome	TIR-ESI	Outcome	TIR-ESI	Outcome	TIR-ESI
Snapshot 1 – Award (2009)	0.461	0.468	0.188	0.092	0.228	0.367	0.563	1.000	1.000	N.A.	C	N.A.	C	N.A.	C_-	N.A.	C
Snapshot 2 – Start of operation (2011)	0.479	0.477	0.188	0.123	0.228	0.367	0.607	1.000	1.000	−1	C	−1	C	−1	C_-	0	C
Snapshot 3 – Reporting time (2015)	0.481	0.514	0.500	0.000	0.228	0.667	0.833	1.000	1.000	−1	N.R.	−1	N.R.	−1	C^+	0	B_{EX}^+
Alternative scenarios																	
Scenario 1 – Award (2009)	0.461	0.468	0.500	0.271	0.228	0.367	0.513	1.000	1.000	N.A.	B_{EX-}	N.A.	B_{EX-}	N.A.	C	N.A.	C^+
Scenario 2 – Reporting time (2015)	0.481	0.514	0.500	0.444	0.228	0.667	0.833	1.000	1.000	−1	N.R.	−1	N.R.	−1	B_{EX}	0	A

Legend: N.A. = not available; N.R. = not relevant

Source: Authors.

Application of the stated measures would also enable a better overall project rating in the award and operating phase.

5.3.3 Concluding remarks

The presented case study of the second phase of motorway Horgos – Novi Sad had a specific project scope (one carriageway) on a relatively exclusive part of the E-75 (M-22) motorway from the Hungarian border to the capital of Serbian province Vojvodina.

The project was awarded and delivered in the period of the global economic crisis after the concession for Horgos – Pozega Toll Motorway had been cancelled. Market conditions in Serbia and the fact that the project was fully financed from the state budget enabled the client to enforce a tight deadline and a specific contract arrangement in terms of payment where 35 per cent of the IPC value for two sections of the project was allowed to be charged at the usual time and the rest was to be paid to the contractor after 30 months.

The analysis conducted herein revealed the weak points of the project and its environment. It was shown that the factors with the highest impact on project performance have originated from project planning and procurement phases and that exogenous market conditions and insufficient institutional capacity and maturity of the public implementing company were especially important for this case and the Serbian context more generally.

Measures that could be applied to improve governance and CSIs and therefore the rating of a project are also identified and presented in the case study ex-ante analysis. The conducted analysis suggested to what degree the potential to achieve better performance can be improved when specific measures are applied. In this direction, work on increasing the Serbian's public sector's capability to properly plan and manage transport infrastructure projects has to be continued.

5.4 The Blanka Tunnel Complex, Czech Republic

Petr Witz

5.4.1 Project delivery

5.4.1.1 Context

The tunnel complex 'Blanka' forms a part of the city of Prague's ring road system. It runs under a heavily urbanized environment right on the edge of the historic city centre. It stretches from Brusnice – to the west of the Prague Castle area – to the municipal district of Troja in the northeast, thus connecting the right and left banks of the River Vltava. The tunnel complex is directly connected to the system of tunnels to the south (Strahovský a Mrázovka) and Břevnovská radial road (connecting the city centre with the Prague Airport). On the right bank of the Vltava, the tunnel connects to Prosecká radial road (exit road to the north) and, in the future, should also connect to the final eastern stretch of the inner-city ring road.

With a total length of 6.4 kilometres, it constitutes the longest road tunnel in the Czech Republic and the longest metropolitan tunnel in Europe. The tunnel complex itself consists of northbound and southbound tubes. It has three main sections and is directly connected to the system of two other inner-city tunnels (Strahovský and Mrázovka), all constructed as parts of the inner-city ring road. The latter is being built in parallel to the so-called Prague ring road (R1) – a motorway that is to encircle the entire city of Prague. The Blanka Tunnel Complex is routed in driven and cut-and-cover tunnels, with connection ramps in key junctions. There are four major intersections along the structure, as shown in Figure 5.5. Finally, a large underground ventilation facility consisting of four engine rooms also constitutes an important feature of the project.

The implementation of the project suffered from a number of political, technical and legal setbacks. Construction started in June 2007, however, in 2008, a series of landslides resulted in fines and new safety and structural require-ments by the State Mining Administration (SMA). In November 2009, it was announced that the opening of the tunnel was postponed by 13 months to 2012. In 2011, it became clear that the original costs estimates would be exceeded by more than 10 bn crowns (approximately 356 M EUR).

In May 2012, the legal uncertainty surrounding extra works and imperfect contracts led the City of Prague to cease payments to contractors. In December of the following year (2013), the contractor Metrostav halted all construction works and an arbitration between the City of Prague and Metrostav began in January 2014. In May 2014, Metrostav resumed construction of the tunnel and ČKD DIZ, the supplier of technology systems, began works. In September 2014, the arbitration ended and the construction of the tunnels was finished. Trojský Bridge opened in October 2014 with a final cost of more than three times

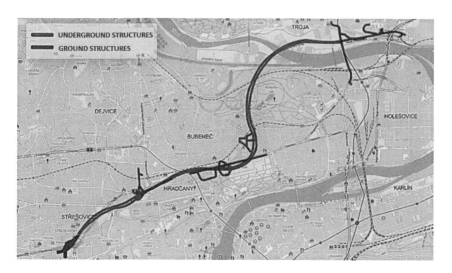

Figure 5.5 Blanka Tunnel complex
Adapted source: Mapy.cz

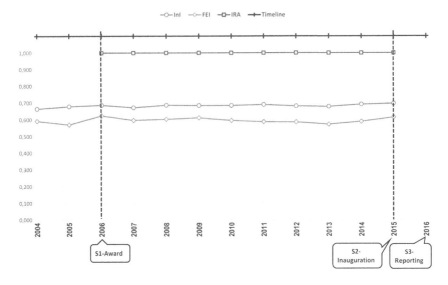

Figure 5.6 Blanka Tunnel project timeline

Source: Authors' compilation.

(1.3 bn Czech crowns – 47.4 M EUR) the originally expected one of 400 million Czech crowns (14.6 M EUR). It should be noted, however, that the design of the bridge changed significantly. The Blanka Tunnel Complex was finally opened to traffic in September 2015. In early 2015, it became evident that the tunnel suffers from technical problems related to dampness, as well as the cabling system.

The tunnel complex was designed to relieve the historic centre of Prague from transit traffic and to speed up transit across the city. The tunnel currently offers faster and uninterrupted connection for vehicles travelling from the west to the east of the city and vice versa. An alternative connection via surface roads remains possible, but the City Hall has introduced certain restrictions at selected points in the city centre to support the use of the tunnel that has become the most convenient way of transit.

Figure 5.6 depicts the project's development (project timeline, IRA, FEI and InI). The project's three snapshots (S1–S3) are also shown, for three key milestones, namely, award, inauguration and one year into operation.

5.4.1.2 Contractual governance

The original idea to build the northern section of the inner-city ring road was laid out in 1993. The plan Blanka was selected among three alternatives. The preparation phase began in 1997 with appropriate changes to the city development plan and by the signing of an agency agreement between the City of Prague and the company IDS that was to become the project and construction supervisor on behalf of the City Hall. In 1998, the project company SATRA was awarded the contract to prepare technical studies and designs for the tunnel complex. Planning decisions and zoning permits were issued in 2003. A lengthy process was concluded with two tenders in 2006. Open call procedures were chosen for both construction and technology parts of the tunnel complex, as with most other important projects in the country, although the number of open calls had been falling over time (Oživení, 2012). The tender for the construction part lasted from the 25 April to the 30 October 2006. Despite the openness of tenders, only three bidders submitted their offers in both calls, with one bidder in each tender disqualified at the very beginning. Metrostav a.s. was announced as the winner of the tender for the construction part on the 26 September 2006 and the contract was signed a month later. The tender for the technology part was launched on the 23 June 2006 and concluded on the 14 May 2007. As a result, the company ČKD DIZ became the supplier of technology systems.

The project was, therefore, based on four separate contracts (design, construction, technology and construction supervision) and their amendments. The division of tasks had been criticised during the process of construction in respect to a series of misunderstandings and disputes among individual contractors and the Prague City Hall. Moreover, and despite, significant imprecisions of original calculations and the resulting significant cost overrun, no strict cap had been incorporated into the original contract for the construction part of the project. Similarly, overruns compared to the original cost expectations were recorded in the contracts for technology and equipment, project design and construction supervision. Evidently, an audit by law firm, White & Case, was published exposing striking examples of ill-prepared project conditions, faulty contracts with insufficient motivation for contractors and mismanagement by the Prague City Hall (White and Case, 2011).

Given the fact that the project was divided into several uncapped contracts under traditional procurement conditions, it is the public authority – the Prague City Hall – that bore most risks related to the construction of the project. There had not been a single instance when the construction supervisor or designer was held responsible for individual failures. Contrary to original expectations, the City of Prague was forced to accept responsibility for the majority of events thatled to substantial cost overruns. The responsibility for operation of the entire tunnel complex also lies with the city holding company since June 2016, although the execution of most tasks during the trial operation period was awarded to a private company.

Based on the above, the GI was 0.313 at project award (2006) and inauguration (2015), increasing to 0.500 at reporting period (2016).

The risk allocation is as depicted in Table 5.6.

5.4.1.3 *Business model efficiency*

PROJECT INTEGRATION

The Blanka Tunnel Complex forms an integral section of the incomplete Prague inner-city ring road, reducing traffic congestion in a number of the city's main

Table 5.6 Blanka Tunnel Complex – ultimate risk allocation

Risks	*Totally private*	⟺		*Totally public*
Design			✓	
Construction			✓	
Maintenance			✓	
Exploitation			✓	
Commercial/revenue			✓	
Financial			✓	
Regulatory			✓	
Force majeure			✓	

Source: Authors.

arteries, while providing faster connection to vehicles travelling between the west and east bank of the Vltava river and also to transit traffic between the northern and western regions of Bohemia.

COMPETENCE OF THE CONTRACTING AUTHORITY

The idea to build the Blanka Tunnel Complex has always been pursued primarily by the representatives of the City of Prague. The Prague City Hall is the main stakeholder in the process. The Blanka Tunnel Complex has been a part of the official policy since the late 1990s.

The City Hall manages the road infrastructure throughout the city – including the Blanka Tunnel Complex – via its company Technická správa komunikací (TSK). In fact, Prague 6 and Prague 7 metropolitan districts, directly affected by the tunnel complex, have also had their say in the process of the tunnel project's planning and construction.

Although there was cross-party support for the realisation of the project, Blanka faced from the outset strong opposition from the Green Party, as well as various environmental groups and activists claiming the promised improvements in Prague traffic are largely illusionary (Auto*Mat, 2009). The Auto*Mat association defending the urban and environmental principles and interests became the most notable opponent and critic of the project trying to offer alternative solutions benefiting public transport, pedestrians and cyclists.

The central government involvement can be seen as very limited and mainly related to changes in regulations that have partly contributed to the increase in the total costs of the project. Among them, additional requirements for safety measures in tunnels by the SMA played a particularly important role. The SMA also fined the contractor for landslides during the construction works.

Most importantly, the Prague City Hall as the contracting authority has been criticised for being defenceless against the numerous price increases and bills for extra works that were uncontested. At several instances, it agreed to arbitrations on disputed issues and lost most of them despite being in the right according to the procurement law (Léko, 2015). What most experts agree on is that the initial and crucial part of the blame lies with the contracting authority and its representatives, who were responsible for setting up the project design, failing to formulate clear specifications for the project and enforce their implementation, signing a largely imperfect contract and failing to manage and coordinate the construction in an informed and responsible way. Frequent political changes and staff reshuffles at the City Hall investment unit are also cited as reasons leading to failure.

Based on the above information (see Table 5.7):

- the CSI was (0.220 at award (2006), decreased to –0.303 at inauguration (2015) and then increased to 0.000 in 2016;
- the RSI remained constant at 0.222 throughout the lifetime of the project.

5.4.1.4 Funding and financing

FUNDING: REMUNERATION SCHEME AND PROJECT REVENUES

Users of the Blanka Tunnel Complex are mostly private users travelling from the west to the east of the capital or the other way around. Public transport vehicles are not expected to use the tunnel. The project is financed from public budget with no other sources of revenues, and hence (see Table 5.7):

- the RAI was 1.000 throughout the project's lifetime;
- the RRI was 0.000 throughout the project's lifetime.

FINANCING SCHEME

The entire project has been financed from the budget of the City of Prague, since it was not eligible for any other sources (EU funds, state budget). At the same time, no debt financing has been used. As a result, the construction of the tunnel complex has had undeniably an immense impact on the budget of the City of Prague and its investment strategy. Over the years, Blanka regularly consumed about half of all investment expenditure. The share has started to decrease only recently with the project being completed (Prague City Hall, 2015). One can argue that committing so much money to the tunnel complex has significantly reduced the space for other investments in the city infrastructure. Blanka took precedence over projects such as the new metro line D or other public transport projects that, according to project critics, would have benefited inhabitants much more.

Based on the above, the FSI was 1.000 throughout the project lifetime.

5.4.1.5 Outcomes

Overall, the project faced serious and unexpected challenges of geological character (three landslides), changes in regulations, flooding and misconduct by both the procuring authority and contractors, as well as prolonged legal battles among the individual parties.

Primarily, the project suffered from the lack of legal clarity that forced the City Hall to halt the payments to contractors on several occasions. The crisis culminated in 2013 with the Mayor of Prague claiming that contracts were not valid from the very start, as they had not been approved by the city assembly.

The tunnel complex also faced criticism for insufficient analysis of its environmental impact. Critics argued that the tunnel would attract even more vehicles to the areas close to the city centre (Auto*Mat, 2011), which in certain cases proved to be true. At the same time, there were claims that the new park areas surrounding the tunnel were designed unprofessionally by engineers reducing green areas.

COST AND TIME TO CONSTRUCTION COMPLETION

The original amount of 21 bn Czech crowns (ca. 766 M EUR) mistakenly presented as the final total cost for the entire project guaranteed not to be exceeded, as claimed by the City Hall representatives (Toman, 2007), soared to the final cost of more than 43 bn Czech crowns (1.56 bn EUR; as of October 2016).

There are several reasons behind the original estimations. First, some items were originally excluded from the calculations of the construction costs by the City Hall representatives even though they constituted indivisible parts of the complex (e.g. key technology equipment or connection to the existing Strahovský tunnel). Also, as the consecutive audits of the project revealed, the figures advertised by the procuring authority at the beginning only included estimated construction costs (White and Case, 2011). Nevertheless, even those were significantly exceeded over the years. The tunnel signalling system was described as excessive consisting of too many colourful lights that required more cabling than necessary. Also, a large part of the inappropriate cable network along the tunnel had to be replaced due to flood damage. Critics claim the damage was only minor in reality and the tunnel complex should have been opened to public regardless (Bělohlav, 2015). As a result of the above, costs spiralled and the date of completion and start of full operation was postponed several times from the original year 2011 to 2016.

ACTUAL VERSUS FORECAST TRAFFIC

The tunnel in the first year of its operation attracted over 30 million vehicles and peaked at about 94.500 vehicles/day, with time savings of an average 44 hours a year (Prague City Hall, 2016), largely in line with original expectations. It became evident that one of the main goals of the project was achieved, that is traffic volumes (in some cases critical over a long period of time before Blanka) on certain major roads in and around the city centre were significantly decreased.

Nevertheless, this partial success has been achieved at the cost of notable traffic increase on peripheral roads leading to the tunnels and on major surrounding intersections, mainly due to the absence of the northern section of the outer city ring road and the remaining eastern part of the inner city circle. As a consequence, certain adjustments to the city transport system had to be made as a result of local demands after the tunnel complex opened to users. It remains to be seen whether these bottlenecks could be alleviated, when the system of the Prague ring road is completed.

Operational costs appear so far not to have diverged much from original estimates. Nevertheless, electricity consumption is considered excessive by some, attributed to intensive lighting in the tunnels.

Finally, a number of the legal disputes are on-going, but in the spring of 2016, the Prague City Hall successfully replaced the ailing company ČKD DIZ responsible for the execution of the trial phase with the city holding company TSK, which in turn contracted the original designer of the tunnel – SATRA – to carry out duties related to its trial operation. Subsequently, a legal action against the City Hall was initiated by ČKD DIZ, claiming additional compensations, which failed.

PROJECT RATING OVER TIME

Table 5.7 presents the results of the ex-post analysis for the Blanka Tunnel Complex project, based on the indicators values, the project outcomes and the TIRESI ratings over time.

The project outcomes in terms of cost and time to completion match the related low TIRESI rating scores that result from the low values of both the Governance Indicator and CSI, the latter decreasing from a very low –0.220 at award further to –0.303 at inauguration. Upon construction completion, both GI and CSI take the average values that are typical for the operation phase of public projects, namely 0.500 and 0.000, respectively. This improvement is reflected in the TIRESI rating for traffic that matches the actual achieved traffic outcome of the project.

5.4.2 *Alternative scenarios of project delivery*

The results of the ex-post analysis of the project revealed that its main limitations comprised of the poor project governance that plagued the project throughout its entire implementation, as well as the lack of capability of the contracting authority together with risk misallocation and technical difficulties, reflected in the low values of GI and CSI, respectively. The above endogenous vulnerability leads to the examination of two alternative scenarios, the first entailing the enhancement of the project governance structure, while the second that of the capability of the contracting authority, both applied at project award.

The procurement method and contractual management – division of the project into four uncapped separate contracts – were seen as key factors behind the unclear distribution of responsibilities, general confusion and subsequent delays and legal disputes. To this end, the first enhancement scenario examines an alternative contractual structure that includes the integration of design and construction services, a shared risk allocation for rising costs, as well as the introduction of performance guarantees and penalties. It should be noted that the replacement of the technology supplier at a later stage in the project by the company that had previously worked on the design indicates that better integration could have been originally achieved.

With regard to the second potential enhancement measure examined, it is evident that the lack of capable, coordinated and decisive leadership on behalf of the Prague City Council impacted project performance in several dimensions. In this case, the enhancement scenario assumes a hypothetical higher and more streamlined expertise of the contracting authority in both project planning and implementation monitoring, while also a shift of the construction risk mostly to the private party.

The first scenario yields somewhat better ratings for the cost and time outcomes, while improves significantly the traffic outcome. The second scenario has no effect on project ratings, indicating the dominant role of the poor governance in the project performance. Nevertheless, a combination of both measures at project award yields the best possible results (see Table 5.7), particularly in terms of the time and traffic outcomes.

5.4.3 Concluding remarks

The Blanka Tunnel Complex has been an extremely challenging project featuring, on the one hand, examples of world-class engineering but, at the same time, suffering from major management issues, legal disputes and force majeure events that led to substantial cost overruns compared to original cost estimates and an almost 5-year delay in completion.

The above casted doubt over the ability of the Prague City Hall to successfully manage large infrastructure projects and have led to several allegations, police investigations and legal battles. It has been argued by auditors that by signing an ill-conceived mandate contract with the project supervisor, the City representatives lost control over the project while failing to motivate the contractor by not including any sanctions or bonuses in the contract. Representatives of the main contractor have blamed the City Hall for insufficient project specifications, lack of leadership and integration and coordination of tasks. According to them, it was the separation of the project into several contracts and lack of clarity over responsibility of the individual contractors that caused the major problems.

Alternative scenarios examined provided evidence that in the absence of the above limitations, there would be high likelihood that the project would not exhibit these severe cost overruns and delay in completion.

5.4.4 Acknowledgments

The contents of this chapter are partly based on the results of the research project SVV No. 260460/2017 of the Faculty of Social Sciences, Charles University in Prague.

Table 5.7 Blanka Tunnel Complex – ex-post analysis and alternative scenarios

Snapshot	Indicator									Cost		Time		Traffic		Revenue	
	FEI	InI	GI	CSI	RSI	RAI	RRI	IRA	FSI	Outcome	TIR-ESI	Outcome	TIR-ESI	Outcome	TIR-ESI	Outcome	TIR-ESI
Snapshot 1 – Award (2006)	0.623	0.686	0.313	−0.220	0.222	1.000	0.000	1.000	1.000	N.A.	B_{EN-}	N.A.	C^+	N.A.	C^+	N.A.	C_-
Snapshot 2 – Inauguration (2015)	0.614	0.696	0.313	−0.303	0.222	1.000	0.000	1.000	1.000	−1	B_{EN-}	−1	C^+	N.A.	C^+	N.A.	C_-
Snapshot 3 – Reporting (2016)	0.620	0.696	0.500	0.000	0.222	1.000	0.000	1.000	1.000	−1	N.R.	−1	N.R.	0	A	0	A^+
Alternative scenarios																	
SC1 – Enhanced governance (2006)	0.623	0.686	0.625	−0.220	0.222	1.000	0.000	1.000	1.000	N.A.	B_{EN}	N.A.	B_{EN-}	N.A.	A	N.A.	A^+
SC2 – Enhanced competence of contracting authority (2006)	0.623	0.686	0.313	0.193	0.222	1.000	0.000	1.000	1.000	N.A.	B_{EN-}	N.A.	C^+	N.A.	C^+	N.A.	C_-
SC3 – Combined enhancements (2006)	0.623	0.686	0.625	0.193	0.222	1.000	0.000	1.000	1.000	N.A.	B_{EN}	N.A.	A_-	N.A.	A	N.A.	A^+

Legend: N.A. = not available; N.R. = not relevant

Source: Authors.

5.5 Reims Tramway, France

Pierre Nouaille

5.5.1 Project delivery

5.5.1.1 Context

This Reims Tramway project, delivered through private (co)financing, included the construction and operation of the first tramway line in Reims (11,2 km, 23 stations and 3 'Park and Ride' areas) and the operation of the metropolitan urban transport system, which also consists of an existing bus line network. The latter was re-organised to accommodate and achieve interoperability with the newly constructed tramway. The restructured bus network consists of two core lines of buses, seven other major bus lines and four bypass lines of buses (Reims Métropole, 2007).

As depicted in Figure 5.7, the tramway line crosses the city from north to south with two routes in the south. It serves residential areas in the north, the central station, the city centre, the hospital area and part of the university. The second route links the central station to a TGV station in the south.

Along the tramway line, the urban space has been improved with the reno-vation of streets, the improvement of public spaces and a new modal space split for the benefit of pedestrians and cyclists. The construction of the project began in May 2008 and the tramway was put into service in April 2011. It should be noted that at the commissioning date, there has been an increase in supply (in commercial kilometres) of almost 6 per cent.

Figure 5.8 presents the Reims Tramway project's development (project time-line, IRA, FEI, and InI). The graph also depicts the project's four snapshots (S1–S4).

5.5.1.2 Contractual governance

Both technical and financial reasons justified the choice of the concession model for this project (Ville and Transport, 2008). These included:

- the need to address certain difficulties of the public transport system with respect to low commercial speeds, stability of usage and increased operating costs;
- avoidance of debt for the urban public transport organizing authority as funding would be supplied by the concessioner;
- improved management of financial resources by the organizing authority with regard to public subsidies for urban transport and the spread of expenditure over time;
- transfer of project management responsibilities to the concessioner, while the organising authority employs a small team for the function;

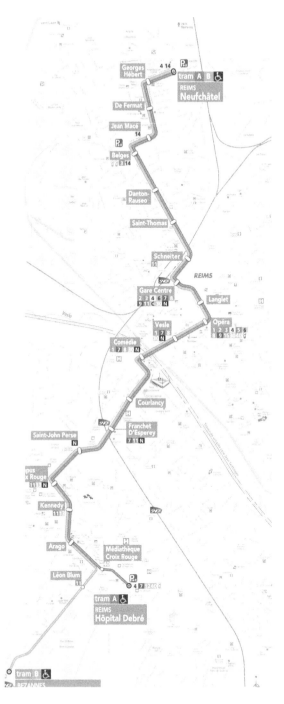

Figure 5.7 Reims Tramway line
Source: CITURA 2016/2017

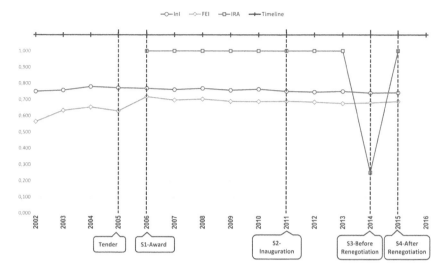

Figure 5.8 Reims Tramway project timeline
Source: Authors' compilation.

- economic optimization of the project;
- compliance with anticipated costs and timelines.

In light of the above, the public transport authority Reims Metropole initiated the development of the tramway infrastructure and decided to use a concession model for public works and public service.

The tendering process was initiated in March 2005 and conducted in accordance with the public service delegation law and regulations, as follows:

- nomination of candidates admitted to the tender;
- receipt of tenders;
- negotiations with candidates;
- delivery of a final best offer by each candidate;
- negotiations with the preferred candidate.

Three expressions of interest were received and all three candidates were allowed to tender for the contract (Ville and Transports, 2008). Hearings and negotiations were conducted with all three candidates. All three submitted an offer. Negotiations were finally concluded with the potential concessioner. The duration of the entire process was almost 17 months.

The contract is a delegation of public service through a concession for public works and public service (Act No. 93–122 of 29 January 1993: Sapin Act). The contract covers the construction of the tram line and related equipment, and operation of the overall urban transport network (tram and bus). It also involves operations related to the urban integration of the tram and the maintenance and renewal of assets allocated to public service. With respect to contractual obligations, the private sector is responsible for:

- design, financing, building, operation and maintenance of the tramway, including passenger stations and park and ride facilities;
- design, financing and development of links with the tramway, in particular those related to urban integration of the tram, according to the conditions and limitations specified in the contract;
- operation of the overall urban transport network (tram and bus). Between 2008 and the implementation of the tramway, operations concerned the existing bus network only;
- maintenance and renewal of assets allocated to the public service. Finally, at the end of the concession period, the project is to be transferred back to Reims Metropole.

The contract defines the obligations of each party in terms of design, construction, maintenance, operation and financial aspects, as described in the following. Risk allocation is depicted in Table 5.8.

Design and construction risks are taken mostly by the concessioner, but the archaeological risk is capped by Reims Metropole. The concessioner is responsible for the maintenance of all property allocated to the public service. In the initial contract, operating risk is mostly assumed by the concessioner.

Commercial/revenue risk is mostly taken by the concessioner that receives the traffic revenue. The annual contribution from Reims Metropole has been defined in the contract, based on operating expenses forecasts and the expected commercial revenue (with a defined ticket price). Risks for interest rates and other financial parameters are mostly carried by the concessioner.

The concessioner assumes the risk of regulations known before the signing of the contract. Reims Metropole takes the risk on regulations that may enter into force after the signing of the contract that were unknown at that date.

Finally, the contract includes conditions for the re-assessment of its financial status. The reassessment takes place three years after the beginning of the

Table 5.8 Reims Tramway – risk allocation

Risks	Totally private	$\Longleftrightarrow$		Totally public
Design and construction		✓		
Maintenance		✓		
Exploitation		✓		
Commercial/revenue		✓		
Financial		✓		
Regulatory			✓	
Force majeure				✓

Source: Author.

operation, then every five years, as well as in particular situations stipulated in the contract.

However, despite the rigorous candidate selection process, the high level of involvement of these candidates in the project design and contractual structure, Reims Metropole was not able to apply some of the contractual provisions due to its lack of knowledge of ridership and network performance. To this end, the contract was renegotiated in February 2015. The planned funding scheme proved insufficient for the network operator to achieve financial equilibrium because of operational difficulties and inaccurate forecasts: operating costs were under-valued and ridership on the network was overestimated. The renegotiation of 2015 attempted to remedy this situation. As a consequence, the level of risk related to fare revenue was adjusted upward, that is, the public transport authority took over part of the commercial risk borne initially by the private consortium.

Based on the above, the GI (see Table 5.9), which was initially considerably high (GI: 0.938), gradually dropped during project implementation and reached a marginal value (GI: 0.500) after the renegotiation of the contract.

5.5.1.3 Business model efficiency

PROJECT INTEGRATION

The tram is in full competition with other transport modes as there are multiple other means of traveling through the city centre. Nevertheless, the competition is limited by the fact that the operator of the tram is granted exclusivity for urban public transport.

COMPETENCE OF THE CONTRACTING AUTHORITY

Since the introduction of the Domestic Transport Orientation Law of 30 December 1982, municipalities or inter-communal structures are responsible for the development and the operation of urban public transport networks. Delegation to private partner is allowed. In this case, Reims Metropole, an inter-communality, is in charge of the development and the operation of the urban public transport network and also acts as the urban public transport authority.

COMPETENCE OF THE PRIVATE SECTOR ACTOR(S)

The concession contract was awarded to MARS (Mobility Agglomeration RémoiSe) that was responsible for:

- the construction of the new tramway line and related equipment;
- the operation of the overall urban transport network (tram and bus) for 30 years from the commissioning date;
- the maintenance and renewal of goods destined for their missions of public service.

MARS is a fully private joint stock company. Its members in 2011 were (see Figure 5.9):

- Caisse des Dépôts et consignations: 30 per cent
- Alstom Transport: 17 per cent
- Transdev: 17 per cent
- Caisse D' Epargne: 8.5 per cent
- FIDEPP: 8.5 per cent
- Bouygues TP: 4.25 per cent
- Colas: 8.5 per cent
- Pingat Ingénrierie SNC Lavalin 2 per cent
- Quille: 2.125 per cent
- Pertuy: 2.125 per cent.

The sharing of responsibilities between shareholders is:

- construction and maintenance: Alstom, Bouygues Travaux Publics, Quille SA, Pertuy Construction and Colas;
- operation: Transdev;
- finance: Caisse des Dépôts et Consignations, Caisse d' Epargne Champagne Ardenne and Natixis.

The CSI values were found to be abnormally low, ranging from –0.018 to 0.000. This highlights one important feature of the project: while most of the construction and operation risks were borne by the private partner, the shareholders

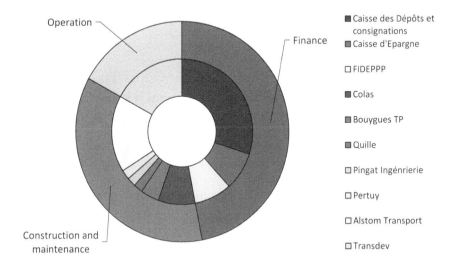

Figure 5.9 Share capital of MARS in 2011
Source: Cerema.

responsible for construction and operation were a minority in the MARS consortium (36 and 17 per cent, respectively). While in retrospect, this fact does not seem to have had any impact during the construction phase, it was a cause of communication problems between the operator and the public transport authority. In fact, direct relationships between operator and public transport authority were not allowed by the contract. These discussions had to take place through the MARS consortium, in which the operator was not necessarily in a position of strength. Taking this fact into account in the estimation of the CSI was, therefore, essential.

The low RSI value (lowest RSI: 0.117, see Table 5.10) reveals a trend that can be generalised for all the urban public transport projects analysed. These projects have a low diversity of income sources apart from fare revenues. In the case of the Reims network, only advertising revenues complement those from ridership. Its share was, however, very limited (a maximum of 4 per cent of network operating costs between 2011 and 2015). In addition, the project is only a part of the urban transit network and does not, by default, hold an important position in the network.

5.5.1.4 Funding and financing

FINANCING SCHEME

The cost of the construction part of the concession contract is 413 M EUR (2011 prices). It covers the tramway construction costs (including the rolling stock, the maintenance centre), the costs of the bus network adjustments. MARS is financing the project as follows:

- 24 M EUR in equity;
- 215 M EUR of external funding (2 loans);
- 174 M EUR of capital subsidy.

Reims Metropole, the urban public transport authority, is involved in financing through:

- direct expenses linked to the project (acquisition of property, costs of diverting the networks, residents' compensation, costs of the organizing authority, etc.). This amounts to 27.2 M EUR;
- a capital grant of 174 M EUR in the construction phase provided partly by:
  - a subsidy form the central government;
  - the city of Reims;
  - an increase in transport tax rate from 1 to 1.8 per cent in April 2005.

Based on the above, the FSI (see Table 5.9) remains stable at 0.869, which indicates strong public contribution.

The tramway is expected to serve 65,000 inhabitants of the city of Reims. The remuneration and revenue scheme is mainly based on a mix of user charges (23.4 per cent) and subventions (76.6 per cent). The private consortium is expected to collect, for its own, commercial revenues related to traffic and auxiliary revenues estimated at 14 M EUR (2006 prices) per year from 2011.

The overall operating subsidy (based on the 2011 network mileage and thus including the tram) is 45.8 M EUR per year as of 2014. This latter subsidy is both a contribution to operating expenses and a fixed amount to cover depreciation costs of the equipment. A part of the operating subsidy paid by Reims Metropole is transferred to financial institutions to secure the loans made to MARS.

The values taken by the RAI (initially 0.923 and then gradually dropping to 0.770; see Table 5.10) directly reflect the risk allocation creep and the inability, from a certain point, of the planned remuneration scheme to cover the project operating costs. As a consequence, the level of risk surrounding commercial revenues, initially underestimated, increased sharply. Amendments made during the renegotiation aimed at returning to the original financial equilibrium, but the level of commercial risk remains high, which explains why the indicator did not return to its original value (RRI is initially 0.312 and then drops to 0.253, see Table 5.10).

5.5.1.5 Outcomes

The project was not hampered by any significant difficulty between the design stage and the end of the construction phase. The public enquiry file (Reims Metropole, January 2007) stipulated that the new tramway lines had to be implemented at the end of 2010. In fact, 2011 is considered as a full year in operation. Actually, the tramway line was implemented in April 2011. In comparison with this document, it could be considered that the commissioning date had been postponed for approximately four months. However, this delay has to be scrutinized in relation to the date of contract award (July 2006) and the beginning of construction (May 2008). The public enquiry file is based on initial feasibility studies. So, the first implementation date has to be considered as an estimation. Its correction is a consequence of the feasibility studies' progress rather than an actual delay. In fact, the grant application submitted by Reims Metropole in January 2009 was based on a commissioning date in April 2011. Therefore, the Reims tramway line can be considered as having been delivered on time.

It is difficult to assess the evolution of the construction cost of the tramway during the life of the project. In fact, the cost of the project is difficult to isolate from the cost of the concession (which includes remuneration of the concessioner, interest rates of the two loans incurred, etc.). Two additional clauses were added to the initial contract between MARS and Reims Metropole. These clauses have led to a cost revaluation. The construction cost went from 257.5 M EUR in 2006 (2005 prices) to 266 M EUR in 2008 (2005 prices). The above amounts cover only the concessioner expenditures. The total investment cost in the second additional clause, after adding external costs (land acquisition, moving of underground networks) and adjusting for inflation is 343 M EUR (2011), close to the final cost estimated in the last financing plan: 345 M EUR.

ACTUAL VERSUS FORECAST TRAFFIC

The implementation of the new tramway line progressed with several difficulties. First, since 2011, the operator met significant technical difficulties with the new contactless ticketing system. This technical malfunction has prevented the concessioner (and indirectly Reims Metropole) from knowing precisely the ridership on these two new lines and, more broadly, on the entire public transport network.

However, the financial balance of the concession contract had been based on an increase in ridership of 40 per cent on the entire network between 2005 and 2015, namely a progression from 30 million trips in 2005 to 42 million trips in 2015 (Reims Metropole, 2007). Even if the exact ridership in 2015 is unknown, its increase during the first few years of operation appears to be much lower than expected (in 2012, approximately 34 million trips were estimated on the public transport network).

ACTUAL VERSUS FORECAST REVENUES

The above overestimation in ridership has been followed by a significant impact on commercial revenues and has created a financial shortfall, which was not to be compensated by the evolution of the operating subsidy paid by Reims Metropole. In fact, this constitutes a lump sum subsidy defined in the concession contract. Without a renegotiation, its variations can only come from penalties or incentives linked to the performance of the service. On this issue, it appeared that network performance monitoring indicators designed in the concession contract were too complex to be calculated.

In December 2014, the operator Transdev, a MARS consortium member, has shown a deficit of 12.5 M EUR (L' Union, 2015). The 2015 review clause of the contract rendered a renegotiation possible.

Given that the operating subsidy remained unchanged through the renegotiation, the latter aimed at finding a financing equilibrium for the concessioner. Accordingly, this renegotiation mainly led to:

- a reorganisation of the bus network and more broadly a rearrangement of the number of network kilometres offered (with a reduction of 60,000 kilometres per year);
- a fare increase for the users;

- the decision to undertake several road improvements (e.g. adaptation of traffic signals) in order to increase the commercial bus speed.

Reims Metropole validated the renegotiation conditions in February 2015. These three adaptations are expected to generate an increase in revenues and a decrease in operating expenditures for the operator, enabling the concessioner to reach financial balance without any increase in the operating subsidy.

PROJECT RATING OVER TIME

Table 5.10 presents the results of the ex-post analysis for the Reims Tramway project, based on the indicators values, the project outcomes and the TIRESI ratings over time. Ratings suggest that the project had marginal likelihood of reaching project targets matching the observed project outcomes.

More specifically, the C^+ ratings were assigned for both cost and time to completion. The traffic and revenue outcome was rated C and B_{EN}, respectively, largely reflecting project performance.

5.5.2 *Alternative scenarios of project delivery*

The *ex-ante* approach, which aims at determining the factors for improvement, therefore, seeks to improve network ridership and concessioner revenue (both outcomes being strongly linked), especially by changing the parameters related to the identified low indicators (GI, CSI, RRI and RSI). The aim is then to:

- increase concessioner revenue or, more accurately, enable the concessioner to achieve financial equilibrium;
- bring network ridership closer to these initial objectives.

In terms of revenue, two categories of improvement may be explored:

- diversifying financial sources that will increase the RSI;
- enhancing existing financial sources that will lead to an increase in RAI and RRI, as this will lower the level of risk related to these sources of funding.

Diversifying sources of revenue is, however, a scarcely conceivable option in the field of urban public transport, where new possible financial sources such as land value capture and urban tolls are based on major innovations that are not necessarily within the reach of a network of the size of Reims. Adaptations should, therefore, be directed towards enhancing existing financial sources, made up mainly by fare revenue. It is, therefore, possible:

- to increase network fares;
- to increase network ridership.

Reims Métropole can authorize a fare increase higher than that provided for in the contract. This increase is theoretically possible: the Reims network was, in 2014, one of the three French urban networks with tramway lines that proposed the lowest fares. But this type of increase may have a negative impact on user satisfaction and, ultimately, on ridership. Raising fares was nevertheless one of the measures selected when the contract was renegotiated in 2015.

Increasing ridership of the transport network is based on coordinated action by the operator and the public transport authority. This action first involves setting up an effective network performance monitoring and control system. Malfunctions of the ticketing system as well as the difficulty for Reims Metropole to apply the performance indicators built into the contract have in fact limited the operator's motivation to offer a quality service. This control must nevertheless be supplemented by:

- making the network more attractive by reducing competition between public transport and other means of transport, especially private cars. Reims Metropole may implement measures to reinforce the commercial speed of public transport, and a parking policy restricting car use in the city centre. These improvements can be taken into account in the estimation of the RSI;
- a better balance between the transport offer and people's needs. This improvement requires frequent adjustments that, under a contract running over a long period of time, require in turn good cooperation between the urban public transport authority and its operator. This topic is beyond the scope of the RSI and involves both contractual issues (GI) and issues related to risk sharing and competence of the actors involved in the project (CSI).

It would in effect involve defining a shared risk between operator and public transport authority closer to what was observed a posteriori. It is noted that despite a commercial risk contractually borne by the concessioner, such risk sharing is not effective insofar as Reims Metropole took over part of the risk during the first renegotiation. This discrepancy between theoretical risk sharing and reality is explained by the existence, for urban public transport networks, of a 'political' risk closely linked to the malfunctioning of an urban public transport network. In the case of Reims, this 'political' risk does not allow the real transfer of commercial risk and operating risk to the private partner.

The CSI, which takes into account the match between competency and risk sharing, partly takes this discrepancy also into account. The level of expertise of the MARS consortium members is not sufficient to justify the operating problems. The stakeholder responsible for operation has recognised expertise at national or even international level. A substantial change to the contract could therefore help to balance risk sharing between the public transport authority and the concession holder in terms of the level of risk actually transferable to the authority. These adaptations however greatly impact the structure of the contract in terms of:

- degree of integration: DBOM contracts involve defining how the project will be operated as of the design phase, which is hardly compatible with the permanent need for readjustment of an urban public transport system;
- duration: financial equilibrium of a public transport network can be predicted only over short periods of time (between 5 and 7 years), which is the average duration of delegation contracts in France.

The above contract enforcement will increase the value of the GI, but will not, in this case, help to improve the CSI.

These potential improvements would be applicable at different stages of the project. Two types of adjustments exist, however:

- adjustments that undermine the foundations of the concession contract. Such adjustments may be taken into account only at the design stage of the contract or would require the termination of the current contract and the design of a new one;
- adjustments that can be made during the contract by means of amendments or renegotiation.

The rating of the above scenarios is presented in Table 5.10. The first scenario had no effect on the expected likelihood of outcomes as GI was sufficiently to start with and the changes did not bring about required improvements in the CSI.

Table 5.9 Impact of possible improvements and stages of implementation

	Impacted indicator	Award Scenario 1	Construction completion Scenario 2	Operation or during a potential renegotiation Scenario 3
Enhance monitoring of the project in order to implement the performance clauses of the concession contract	GI CSI	✗		✗
Increase the share of the operator in the social capital of the concessioner	CSI	✗		
Allocate a realistic level of commercial and operational risk to the concessioner	GI CSI	✗		
Increase fare prices	RRI RAI	✗	✗	✗
Enhance the exclusiveness of the transport project	RSI		✗	✗

Source: Author's compilation.

Table 5.10 Reims Tramway – ex-post analysis and alternative scenarios

Snapshot	Indicator										Cost		Time		Traffic		Revenue	
	FEI	InI	GI	CSI	RSI	LOC	RAI	RRI	IRA	FSI	Outcome	TIR-ESI	Outcome	TIR-ESI	Outcome	TIR-ESI	Outcome	TIR-ESI
Reims Contract award – Snapshot 1 (2006)	0.718	0.769	0.938	−0.018	0.151	0.412	0.923	0.312	1.000	0.869	N.A.	C^+	N.A.	C^+	N.A.	C	N.A.	B_{EN}
Reims Inauguration – Snapshot 2 (2011)	0.690	0.751	0.563	−0.018	0.127	0.412	0.923	0.312	1.000	0.869	−1	C_-	−1	C_-	N.A.	C	N.A.	B_{EN}
Reims before renegotiation – Snapshot 3 (2014)	0.679	0.741	0.563	0.000	0.117	0.412	0.770	0.316	0.250	0.869	−1	N.R.	−1	N.R.	−2	C	−1	B_{EN}
Reims after renegotiation – Snapshot 4 (2015)	0.688	0.742	0.500	0.000	0.163	0.412	0.770	0.253	1.000	0.869	−1	N.R.	−1	N.R.	−1	C	−1	B_{EN}
Alternative scenarios																		
Reims Sc1 (2006)	0.718	0.769	0.938	0.056	0.151	0.412	0.847	0.249	1.000	0.869	N.A.	C^+	N.A.	C^+	N.A.	C	N.A.	B_{EN}
Reims Sc2 (2011)	0.690	0.751	0.563	−0.018	0.207	0.588	0.847	0.249	1.000	0.869	−1	B_{EN}	−1	B_{EN}	0	B_{EN-}	0	B_{EN}^+
Reims Sc3 (2015)	0.688	0.742	0.500	0.000	0.223	0.588	0.770	0.194	1.000	0.869	−1	N.R.	−1	N.R.	−1	B_{EN-}	−1	B_{EN}^+

Legend: N.A. = not available; N.R. = not relevant
Source: Authors..

The second scenario seems more efficient. A key differentiation is the support provided to the project in the network (increase in the level of coopetition (LOC), which is included in the RSI).

The final scenario also includes a relative improvement in CSI, but not capable of changing the expected likelihood of outcome achievement.

5.5.3 Concluding remarks

The Reims Tramway is a project with highly vulnerable internal structure, the latter manifesting in the operating phase of the project. The limits concerning questions of cost saving and revenue robustness for this project have penalized the operation of the project as predicted by its TIRESI ratings.

Possible improvements considered in the ex-ante analysis are directly applicable and have already been implemented through the 2015 renegotiation. Other potential ex post enhancement scenarios concern more substantial proposals that challenge the principles of the concession (a long-term contract transferring the majority of the risks to the concessioner). These would be therefore more related to designing a new project rather than improving the existing one.

As shown in Table 5.10, the proposed enhancements lead to minor improvements in all performance outcomes, especially when implemented after project inauguration (2011) and during operation of the tram (2015).

In this approach, the TIRESI has been used to test scenarios used during renegotiations.

Finally, one of the causes of the difficulties encountered by the Reims network is an overestimation of ridership targets. While the form of the contract may partly explain this overestimation, its true cause is to be sought in the initial studies, which is to a large extent outside the scope of the present analysis.

5.6 Brabo 1, Belgium

Eleni Moschouli and Thierry Vanelslander

5.6.1 Project delivery

5.6.1.1 Context

Brabo 1 was the first PPP for public transport in Flanders. The investment size of this project equals 125.8 M EUR (2009). The project involved the design, financing, construction and maintenance of the civil, mechanical and electrical infrastructure associated with two separate tramway extensions in the eastern part of the city of Antwerp: (i) The Antwerp–Deurne section was extended to Wijnegem, (ii) The Antwerp–Mortsel section was extended to Boechout. Additionally, the Brabo 1 project also included an upgrade of the public space in general. Particularly, the project provided for a comprehensive renewal of all associated street infrastructure (including pavements and street furniture) for motor traffic, cyclists and pedestrians. A substantial tram stabling and maintenance depot, located on one of the lines (with office accommodation), was also included. Both trajectories were extended in favour of inhabitants of the Antwerp suburbs. People living in this first 'circle' of municipalities around Antwerp were seeking rapid, punctual tramway connections, since heavy traffic congestion affected their day-to-day commuting to and from the city (Beheersmaatschappij Antwerpen Mobiel, 2013).

The idea of constructing the two tramway extensions in the eastern part of Antwerp city dates back to 2000. Although the Brabo 1 project has served the local interest, it was driven from the Flemish Government level. The region announced the urgent need for improvements in the Master Plan 2020, which was launched in 2000. However, the tendering took place in 2007 and the project was awarded in 2009. More specifically, the contract was approved in August 2009 and the construction works started in October 2009. The first section Deurne–Wijnegem was put in service on February 2012, while the second, Mortsel–Boechout, in August 2012.

Figure 5.10 illustrates the project's development (project timeline, IRA, FEI, InI). The graph also depicts the project's two snapshots at award (2009) and operation (2014).

5.6.1.2 Contractual governance

An open call for the expression of interest was announced in July 2007. Three consortia were selected for negotiations for the DBFM tender, and, on April 2009, submitted their final offers: DANK, THV Silvius and Travant. Subsequently, THV

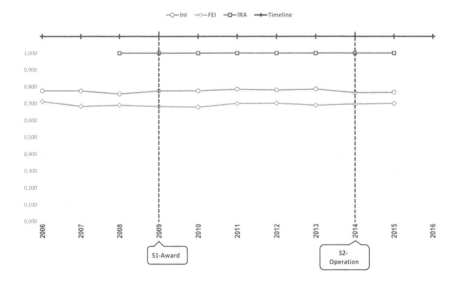

Figure 5.10 Brabo 1 project timeline

Source: Authors' compilation.

Silvius was chosen as the contractor for the Brabo 1 project. The procurement process was concluded within 26 months. The contract duration is 35 years.

The global financial–economic crisis between 2008 and 2012 delayed the procurement process; as bidders had failed to find the necessary external financing for their offers, contractual close and financial close were postponed. A new round for offers had to be arranged by the contracting authorities. The delay ultimately proved to be approximately six months, as the tendering procedure was halted between November 2008 and spring 2009. Further delays were caused by a lawsuit that was filed by one of the bidders (Travant), claiming to be unfairly excluded from the tendering procedure.

The contractual regime was based on a fairly standard form of the DBFM agreement, based on English PFI standards and the Dutch standard DBFM contract of the Dutch Directorate-General for Public Works and Water Management (for an updated version of this contract, see RWS, 2012). The contract for the Brabo 1 project was signed by the existing transport operator De Lijn nv, the Antwerp Agency for Roads and Traffic AWV and the special purpose vehicle (SPV). In this contract, a distinction was made between tramway availability fees and non-tramway availability fees. The former are being paid by De Lijn nv, whereas the latter are paid by AWV. The operation, maintenance and ownership of trams and buses will rest with De Lijn.

The contract can be considered efficient because: (i) it encouraged competition, (ii) there was an integration of design and construction in the services provided, (iii) the contractor had to solely carry the risks of rising costs and (iv) it allowed incentives for performance. Some contract inefficiencies were also observed because (i) revenue risks are not shared but concentrated only on one party (public), (ii) there was not a collective estimation for investments and (iii) the contractor is not obliged to pay a penalty if completion dates are not met. The contract is also flexible because it includes clauses enabling either or both updating of service and price changes and, also, clauses indicating that the client has an option to terminate the agreement prematurely without cause.

Although the public authority has transferred quite a number of responsibilities, it has retained some of its financial activities; it has invested equity in the SPV, and therefore, the public sector shares significant financial risk. The same applies also to the allocation of maintenance risk, which is almost equally shared between the public and the private partner; exploitation and revenue risks are borne entirely by the public sector and, also, regulatory and force majeure risks are mostly borne by the public sector. The design and construction risks are almost entirely allocated to the private partner (see Table 5.11).

Based on the above, the GI was 0.688 throughout the project lifetime.

5.6.1.3 *Business model efficiency*

PROJECT INTEGRATION

The Brabo 1 is located in an urban area with multiple transportation alternatives for users (bus services, cycling). Brabo 1 connects to train stations, bus stations

Table 5.11 Brabo 1 – risk allocation

Risks	Totally private			⟷		Totally public
Design and construction		✓				
Maintenance			✓			
Exploitation						✓
Commercial/revenue						✓
Financial			✓			
Regulatory					✓	
Force majeure				✓		

Source: Authors.

and park-and-ride facilities (physical integration). The project is well connected with the bus lines, which are the alternative public transport means in the city of Antwerp and a uniform ticket fare allows transit from the tram to the bus service for 60 minutes (operational integration). De Lijn is the autonomous public authority that manages/operates both the tramway services and the bus services in the entire Flemish region (authority integration). The project is also integrated into the transport and other planning policies (policy integration). The network integration influences somewhat positively the project with respect to control over demand.

COMPETENCE OF THE CONTRACTING AUTHORITY

De Lijn nv and the Antwerp Agency for Roads and Traffic AWV are the two contracting authorities in this project. AWV is involved due to its responsibility for the renovation of roads, which was also part of the Brabo 1 project. The two authorities jointly signed a DBFM contract with the SPV. Notably, as this was the first PPP tramway-related project in Flanders, neither authority had extensive experience.

In general, the project was well planned. There was a clear policy with respect to the project, a political decision to adopt PPP and no inaccurate pre-project information identified, thus revealing the experience of the Contracting Authority in planning Brabo 1. However, during the project implementation phase, some issues were raised. An ex-ante evaluation in order to examine whether the PPP would be cheaper than traditional procurement was not carried out. A respective evaluation was made after deciding to deliver the project as a PPP.

De Lijn not only plans the infrastructure, but also monitors its implementation (construction) and is also the operator. Regarding monitoring, the contracting authority has a good project management record and capable staff to monitor the project. It is also worth mentioning that there were no lengthy renegotiations and, in general, the project received positive press reviews and had the support from various stakeholders. Last but not least, the Contracting Authority is a highly experienced operator.

Project sponsors are Lijninvest nv (equity) and DG Infra+ (equity). The latter is a Belgian investment fund, established by GIMV and Dexia. Four banks provided senior debt: KBC, Dexia Credit Local, Dexia Bank Belgium and the Bank of Dutch Municipalities (Bank Nederlandse Gemeenten). The SPV is Project Brabo 1 nv (Lijninvest nv (24 per cent), Beheersmaatschappij Antwerpen Mobiel (24 per cent) and THV Silvius (52 per cent)). THV Silvius was also the consortium of subcontractors: DG Infra+ (investment fund), Heijmans Infra nv (construction company), FrateurDe Pourcq nv (construction company) and Franki Construct nv (construction company). All the companies of the construction consortium (THV Silvius, 52 per cent) are top national players with a sufficient ability to construct and, therefore, the construction risk was appropriately allocated to the constructors. Innovation was adopted and successfully applied, which also mirrors the competence of the concessioner.

Based on the above, and in combination with the risk allocation structure, as presented in the previous section:

- the CSI was 0.363 at award (2009) and decreased to 0.000 in 2014 (operational phase);
- the RSI remained stable at 0.142 at award (2009) and in 2014 (operational phase). The low level of project exclusivity has an impact on the level of coopetition, which reduces significantly the RSI values. Similarly, the business scope of the infrastructure (business servicer) and the absence of non-transport business activities within the project, which could bring revenues, contribute also to a low RSI value (Table 5.12).

5.6.1.4 *Funding and financing*

The Brabo 1 project is available to private (passenger) traffic. The remuneration scheme is availability fees. Project revenues are generated by passenger fares that are collected by the operator (De Lijn nv).

Based on the above:

- the value of the RAI indicator is equal to 0.667 and remained unchanged in both snapshots (award: 2009 and operation: 2014).
- the RRI is equal to 0.308 for both snapshots. Notably, passenger fares are assumed to correspond to 30 per cent cost coverage.

Equity covered 10 per cent of the total financing needs, of which Silvius (Tinc comm.v.) provided 52 per cent, and Lijninvest and BAM 24 per cent each. The size of the equity capital was 4.6 M EUR pure equity and 13.8 M EUR subordinated shareholder loan. Financing through debt (senior debt) was provided by the following Commercial banks: KBC, Bank Nederlandse Gemeenten (Bank of Dutch Municipalities), Belfius and Caisse des Dépôts et Consignations, which together covered the remaining 90 per cent of the total financing cost. The size of debt capital was 161.7 M EUR and the total capital was 180.1 M EUR. More specifically, the Leading bank (KBC Bank) provided debt finance equal to 53.9 M EUR senior debt, while the other banks such as the Dexia Credit Local, Belfius Bank and Bank Nederlandse Gemeenten provided debt finance equal to 107.8 M EUR senior debt.

Based on the above, the FSI was 0.720 at award (2009) and remained stable until its reported operational phase (2014) (see Table 5.12). The financing scheme used is a low-cost financing scheme.

5.6.1.5 *Outcomes*

Brabo 1 is a successful project that achieved the time, traffic and revenues targets, as presented next. No renegotiations took place. With respect to the availability and reliability of the project, they were fully in line with the expectations (100 per cent). The same applies to the maintenance costs, the safety and security (security incidents are within expected range). Regarding the economic, social, environmental and institutional impacts of the project, they were as expected.

COST AND TIME TO CONSTRUCTION COMPLETION

Brabo 1 was delivered ahead of schedule. However, the project faced a limited cost overrun (1 per cent increased construction costs) because of the design modifications introduced by AWV and De Lijn. These modifications included inter alia (i) the construction of an extra traction station, (ii) a different planning of the construction in a shopping street and (iii) the addition of a junction to the project configuration.

ACTUAL VERSUS FORECAST TRAFFIC

The actual traffic of the project was as forecast.

ACTUAL VERSUS FORECAST REVENUES

Actual revenues were also as predicted.

PROJECT RATING OVER TIME

Table 5.12 summarises the indicator values results of the Brabo 1 project and provides the respective TIRESI ratings over time. Based on the values of the indicators in Table 5.12, the project appeared to have a low to average likelihood of achieving the cost and time to completion outcome (B_{EN}), as well as revenue forecast (B_{EN}^+ and B_{EN}) and an even lower potential of reaching its traffic forecast (C^+ and C). Overall, it may be concluded that the project performed better than expected.

5.6.2 Alternative scenarios of project delivery

Project improvement scenarios may be considered during all project phases (Voordijk *et al.*, 2016, p. 31). Therefore, improved scenarios will be considered for both project phases of the Brabo 1 project, for which data are available; award phase (2009) and operational phase (2014). The indicators that will be re-examined are the GI and RSI, because these are found as significant for the achievement of the cost to completion outcome (see Roumboutsos *et al.*, 2016).

The GI's actual value is 0.688, which is slightly lower than the minimum threshold defined (0.700 for urban transit projects). However, the following action is suggested in order to improve governance and therefore the likelihood of achieving the cost to completion outcome: *obliging contractors to pay a penalty if completion dates are not met*. Notably, this scenario also tests the level of fuzziness related to the indicator threshold values.

The implementation of this action could have been feasible in the early stages of the project, when the contract was created. The GI increased from 0.688 to 0.750. Lastly, an improvement measure is suggested for the RSI.

Although a high value of the FEI (0.696) is supportive of a high RSI, because implementation context conditions would allow for business initiatives, this is not the case for Brabo 1. The RSI has a low value equal to 0.142. The following improvement measure is suggested: *attracting/increasing non-transport revenues*.

The implementation of this action is deemed realistic and feasible. Evidently, there is very limited margin for RSI to improve, because even after the application of the above measure, the indicator increased only by 0.017 (RSI = 0.159).

Following the implementation of the above potential enhancement scenarios, the TIRESI ratings for the project were significantly improved for the time and cost to completion outcome (B_{EN} to A_) for the first scenario-award phase but no significant change is reported for the second scenario.

5.6.3 Concluding remarks

The Brabo project is a Belgian PPP tram project that was awarded and started to be constructed in 2009, when the financial crisis was gaining momentum. However, this did not affect critically the project, since such urban transit projects are not directly affected by the financial-economic context of the country where they are located. In general, it is a successful project because it achieved all its performance targets apart from one, the cost to completion. However, cost overruns were only a mere 1 per cent.

Table 5.12 Brabo 1 – ex-post analysis and alternative scenarios

Ex post analysis

Snapshot	Indicator									Cost		Time		Traffic		Revenue	
	FEI	InI	GI	CSI	RSI	RAI	RRI	IRA	FSI	Out-come	TIR-ESI	Out-come	TIR-ESI	Out-come	TIR-ESI	Out-come	TIR-ESI
Snapshot 1 – Award (2009)	0.682	0.775	0.688	0.363	0.142	0.667	0.308	1.000	0.720	N.A.	B_{EN}	N.A.	B_{EN}	N.A.	C^+	N.A.	B_{EN}^+
Snapshot 2 – Operation (2014)	0.696	0.764	0.688	0.000	0.142	0.667	0.308	1.000	0.720	–1	N.R.	1	N.R.	0	C	0	B_{EN}
Alternative Scenarios																	
Scenario 1 – Award (2009) improved	0.682	0.775	0.750	0.363	0.159	0.667	0.308	1.000	0.720	N.A.	A_-	N.A.	A_-	N.A.	C^+	N.A.	B_{EN}^+
Scenario 2 – Operation (2014) improved	0.696	0.764	0.750	0.000	0.159	0.667	0.308	1.000	0.720	–1	N.R.	1	N.R.	0	C	0	B_{EN}

Legend: N.A. = not available; N.R. = not relevant
Source: Authors.

Improvement measures were suggested in order to test if these would increase the likelihood of the cost target to be achieved. These measures increased the GI from 0.688 to 0.750 and the RSI from 0.142 to 0.159. As a result, the rating of the project was improved from B_{EN} to A_- at the award phase providing a better estimation of the respective outcome and from C_- to C^+ at the operational phase. This, also, denotes that indicator threshold values should be addressed as a range (fuzzy value) and ratings should be considered for both below and above the threshold value.

5.7 E39 Klett-Bardshaug Road, Norway

Alenka Temeljotov Salaj, Iosif Karousos and Panayota Moraiti

5.7.1 Project delivery

5.7.1.1 Context

E39 Klett-Bardshaug in Sør-Trøndelag County is one of first three PPP transport infrastructure projects in Norway (the other two being E39 Lyngdal – Flekkefjord and E18 Grimstad – Kristiansand), which were chosen to be initiated in 2000 and were included in the National Transport Program (NTP) 2002–2011 (Bjorberg *et al.* 2014; Odeck 2014). The idea of starting with PPP projects was based on the Government's plans to prioritise resources. There were several incentives for PPP projects stipulated in the national strategy, such as financial benefits, predictable financing, early start-up, long term planning, no main- tenance lag, less bureaucracy, stimulating innovation, risk sharing, professional engineering. Nevertheless, the main reasons for choosing the concession model in the E39 Klett-Bardshaug project were to allocate roles and responsibilities according to the principles of the national road administration and to allocate risk where it could be most efficiently handled (Bjorberg *et al.* 2014). The Norwegian Parliament (Stortinget) wanted to test if the PPP model is a more efficient way of developing road projects than traditional procurement and also whether PPP allows political influence and public control.

In the region, E39 is one of the most important roads. Businesses located along the road generate high transport demand and operate in sectors such as oil and gas, industry, fish farming and agriculture. An increasing number of these operate internationally and need efficient access to air transport. The nearest airport terminal is Vaernes, approximately 25 kilometres north of Trondheim. The old road had sharp curves, poor visibility, and few possibilities for overtaking. Over a three-year period, there were more than 80 incidents with personal injuries and four deaths. The traffic growth in the county was approx. 1.5 per cent annually, whereas the actual road traffic had a growth rate of 2.5 per cent per annum.

The project was conceived in 2002, the announcement of pre-qualification and distribution of tender documents was made in 2002, evaluation and negoti- ations were on-going until the end of 2002, and the contract was awarded in the beginning of 2003. The road was opened in 2005, two months ahead of schedule.

Figure 5.11 E39 Klett-Bardshaug Motorway
Source: Kjell Herskedal.

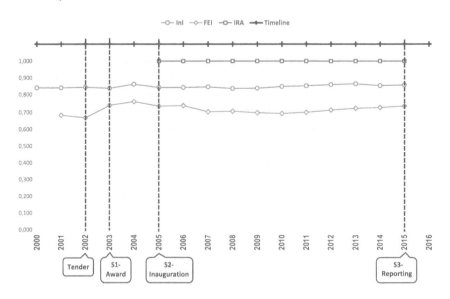

Figure 5.12 E39 Klett-Bardshaug Motorway project timeline
Source: Authors' compilation.

The project included technical innovation, project management innovation and financial innovation. During its operation, the project has managed to reduce the number of accidents and improve the technical conditions of the road (Bjorberg *et al.* 2014), while being sufficiently profitable for the SPV (Orkdalsvegen AS) to pay dividends to its owners.

Figure 5.12 illustrates the project's development (project timeline, IRA, FEI, and InI).

5.7.1.2 *Contractual governance*

In October 2001, the tender was announced nationally and internationally in the *Official Journal of the European Union* (OJEU). The Norwegian Public Roads Administration emphasised the need to identify candidates with the necessary knowledge, experience, capacity and financial strength to complete a project of this type, magnitude and complexity, which were the requirements in the pre-qualification application. Tender documents were distributed in February 2002. On the basis of the received pre-qualification applications, the Norwegian Public Roads Administration selected four candidates (the stated maximum) who were invited to submit a tender by July 2012, and potentially negotiate the terms for entering the PPP contract. Upon receipt of tenders, three bidders were selected for negotiations in the early stages, and two bidders went ahead to the final stage of the negotiations. The evaluation and negotiation process lasted from July until December 2002. The Norwegian Public Roads Administration decided to include a best and final offer (BAFO) stage in order to award the contract to the candidate with the most economically advantageous offer considering the complete set of evaluation factors presented in the tender documents. Contract assignment period was from December 2002 until January 2003.

The decision was made not to transfer the traffic revenue risk to the PPP Company. The concessioner assumes full responsibility for design, construction, financing and operation of the road. The operation period is 25 years after the construction is completed. The PPP Company has general responsibility for the availability and safety of the road, and for environmental and aesthetic standards. Details relating to functional requirements, performance and quality standards, which ensure the public authority's overall objectives are met, are specified in the contact. The project was finished in 2005, without any delays.

The concessioner's remuneration is based on the road's availability and quality, emphasising increased safety with a bonus, incorporated in the contract related to reduction in the frequency of accidents. The performance of the private contractor in operating is also measured by monitoring the operation of key systems on the road such as lighting, air quality and safety systems. The state is responsible for traffic risk in the project and traffic forecasts have been estimated. The PPP Company may receive a compensation for wear of the road if HGV (heavy goods vehicle) traffic exceeds a certain level above the specified traffic forecasts.

Risk allocation is illustrated in detail Table 5.13. In addition, the private sector is also responsible for the following risks and respective costs:

- cost of compliance with all general changes in legislation and statutory requirements;
- changes in scope of service specifications for operations and maintenance by the SPV;
- cost overruns for operations, maintenance and service specifications;
- damage to infrastructure;
- latent defects on both road sections (old and new);
- adverse weather conditions.

The public sector is responsible for changes in the scope of service specifications by the Norwegian Public Roads Administration during the operation stage.

Finally, both parties share the cost of latent defects beyond the end of the contract. In general, the state bears the risk for pre-contract award issues, whereas the PPP Company bears the risk for post-contract award preparations. The risk elements concerning land acquisition are complex and have been partly defined as a joint responsibility of the parties. The risk elements in design, financing, construction and operation lie mainly with the contractor; however, there are elements that the State can more suitably handle, and these are allocated to the public sector.

Based on the above, the GI was 0.563 throughout the lifetime of the project.

5.7.1.3 Business model efficiency

PROJECT INTEGRATION

The E39 Klett-Bardshaug is a unique project offering:

- efficient access to Vaernes airport;
- reduction of traffic congestion in the main arteries to Trondheim;
- improvement of a section of the E39, a 1330-kilometre long north-south road in Norway and Denmark;

Table 5.13 E39 Klett-Bardshaug Road – risk allocation

Risks	Totally private			Totally public
Design and construction		✓		
Maintenance		✓		
Exploitation		✓		
Commercial/revenue				✓
Financial		✓		
Regulatory			✓	
Force majeure				✓

Source: Authors.

- improvement of a mostly two-lane undivided road, with relatively short sections of motorways or semi-motorways.

In Norway, infrastructure policy is published in the Norwegian NTP, which outlines how the Government plans to prioritise resources. NTP has a duration of ten years, but is revised every fourth year. The national agencies Norwegian Air Traffic Authority (Avinor AS), Norwegian Coastal Administration (Kystverket), Norwegian National Rail Administration (Jernbaneverket) and Norwegian Public Roads Administration (Statens vegvesen) are responsible for air, sea, rail and road transport in Norway, respectively (The Norwegian National Transport Plan, 2013). The seaports are owned locally or inter-municipally, but administered by the government (NAF, 2015a).

The Norwegian Public Roads Administration is responsible for the planning, construction and operation of the national and county road networks, vehicle inspection requirements, driver training and licensing. It also grants subsidies for ferry operations. On matters pertaining to national roads, such as the E39 highway, the Public Roads Administration is under the direction of the Ministry of Transport and Communications. The objective of the Norwegian Public Roads Administration is to develop and maintain a safe, eco-friendly and efficient transport system.

There is no official standardised contract for PPP projects in Norway, but private law firm Schjodt has cooperated with the National Public Roads Administration to develop a standard PPP contract. For the three PPP transport projects, including E39 Klett-Bardshaug road, a Design, Build, Finance, Operate (DBFO) contract model was designed. The contractual parties are the Norwegian Public Roads Administration and a PPP company.

However, it should be noted that this was one of the first PPP projects in Norway and the contracting authority had limited experience.

The PPP Company, Orkdalsvegen AS, is a SPV formed jointly between Skanska Infrastructure AV and John Laing Infrastructure Ltd. Skanska Norway is a key subcontractor with full responsibility for the operation of the road. The project also provides the opportunity for project sponsors to demonstrate their skills.

Based on the above, and in combination with the risk allocation structure, as presented in the previous section (see Table 5.14):

- the CSI was 0.722 at award (2003), decreased to 0.556 at inauguration (2005) and then increased again to 0.667 in 2014;
- the RSI was 0.200 at award (2003), inauguration (2005) and reporting period (2014).

5.7.1.4 Funding and financing

FUNDING: REMUNERATION SCHEME AND PROJECT REVENUES

The motorway is open to all private vehicles and freight traffic. E39 Klett-Bardshaug serves, however, passenger cars for commuting purposes, travelling to Vaernas airport, logistic purposes and accessing different travel destinations and roads down the E39 national road.

Orkdalsvegen's remuneration is based on: payments based on the road's availability and quality, payments for operation and maintenance (including road monitoring), payment for complying with the output specifications of road delivery (friction, visibility of signposts, air quality in tunnels, winter maintenance, etc.), a safety bonus payment linked to the number and severity of accidents and compensation received for unexpected traffic volumes.

Availability payments are a measure of the availability of the road to the traveling public, and reward the private contractor for designing a high-quality road that needs little maintenance, and ensuring that the maintenance is programmed to avoid busy and congested times. The payment is linked to the output specifications for the road, including the required standards of performance and its operation. Availability payments are divided between different sections of the road, and standards are set for road conditions (e.g. skid resistance) below which the road will be deemed to be non-available.

Payments for operations and maintenance: the performance of the private contractor in operating the road is measured in several ways. One such measure is the operational performance and maintenance of the road, which includes the cleanliness of the road and surrounding areas, road surface tests, and the time taken to replace broken assets (e.g. traffic signs), clear snow or salt the road. These measures ensure that the operation, design and maintenance of the road are kept up to a high standard. The performance of the private contractor in operating the E39 road is also measured by monitoring the operation of key systems on the road such as lighting, air quality and safety systems.

Safety payments: it is a key objective of the Norwegian Public Roads Administration that the safety record of the E39 road should be of a high standard. A share of the payments for the new road is, therefore, linked to the annual number of personal injury accidents and how this relates to pre-determined targets levels. The target level for the Norwegian Public Roads Administration is set by reference to experience on other similar roads and to Norwegian road safety targets. It is envisaged that the safety payment will be a 'bonus' payment to the contractor for improved safety performance on the E39 road.

Traffic payments: the State is responsible for the traffic risk of the project and traffic forecasts have been developed. The PPP Company may receive compensation for wear of the road if HGV traffic exceeds a certain level above the specified traffic forecasts.

Based on the above (see Table 5.14):

- the RAI was 0.333 in all snapshots (from award in 2003 to the reporting period in 2014);
- the RRI was 0.667 in all snapshots (from award in 2003 to the reporting period in 2014).

FINANCING SCHEME

The project was 100 per cent financed by Orkdalsvegen AS. The financing structure is a consortium managed by the Norwegian financial institution Nordea (1 per cent equity, 9 per cent subordinated loan, 90 per cent bank loans). Skanska and Laing Roads Ltd. invested each around NOK 73 M (9 M EUR).

The up-front (2001) estimated value of the construction element of the contract was NOK 1000–1200 M (approximately 125–150 M EUR). The operating and maintenance costs were estimated at NOK 16 M (2 M EUR) per year.

The construction was financed as follows:

- equity: 9 M EUR;
- bank loan: 200 M EUR.

Notably, the project reached financial close in 2005.
Based on the above, the FSI was 0.719 in all snapshots.

5.7.1.5 *Outcomes*

E39 Klett-Bardshaug road is a successful project and has exceeded forecast demand.

The project has shown to reduce the number of accidents and improve the technical condition of the road. This is in line with the state's requirements and the objectives defined in the contract. Furthermore, the project is sufficiently profitable for the SPV to pay dividends to its owners. So far, the project has met the main objectives of both contractual parties.

COST AND TIME TO CONSTRUCTION COMPLETION

E39 Klett-Bardshaug road was built on time and on budget. In comparison with the classic procurement tenders, the construction was efficient without any serious conflicts, while the construction time was reduced to 60 per cent.

ACTUAL VERSUS FORECAST TRAFFIC

The motorway was fully operational as of 2005. The E39 Klett-Bardshaug had an average traffic ranging from 5600 to 8700 vehicles per day in 2000. In 2015, the average traffic was approximately 9000 vehicles per day according to the operator.

ACTUAL VERSUS FORECAST REVENUES

The Norwegian Public Roads Administration aims to increase safety on the accident-prone route. A bonus is therefore also issued if the frequency of accidents is reduced. The operating period started when the road was opened, and the PPP Company is paid an annual compensation during this period. The payment is based on incentives and performance against a number of pre-defined criteria incorporated into the payment mechanism of the PPP contract.

PROJECT RATING OVER TIME

Table 5.14 presents the results of the ex-post analysis for the E39 Klett-Bardshaug project, based on the indicator values, the project outcomes and the TIRESI ratings over time. It is evident that the positive project outcomes match the TIRESI rating scores for all case snapshots, predicting the high likelihood of success for the motorway.

5.7.2 Alternative scenarios of project delivery

The E39 was structured as a very robust project, developed as a pilot to test the PPP delivery model in the country, as presented in Part 1 of the case analysis. To this end, no significant improvements can be achieved with respect to the project endogenous indicators.

In light of the above, the ex-post analysis examines two alternative scenarios: one that tests a lower FSI value for the award year and a second one that involves a hypothetical unfavourable implementation context, where the FEI drops to values below 0.60.

As shown in Table 5.14, the first scenario entails a higher risk financing structure, which is more typical in a PPP delivery model for infrastructure projects and renders the FSI = 0.506. This change does not affect the high TIRESI rating for all project outcomes, which maintain their 'A' or 'A$^+$' scores. The second scenario yields only slightly reduced rating scores, whereby traffic and revenue outcomes maintain their positive 'A' rating, providing strong evidence of the project's resilient internal structure even to adverse implementation conditions.

5.7.3 Concluding remarks

The E39 Klett-Bardshaug is an overall successful project with positive performance in terms of all four outcomes (cost, time, traffic, revenue). This success level is preserved even in the case when a PPP scheme incorporates significantly higher risk financing sources. Besides this, the favourable implementation context (high InI and FEI values) definitely contributes to the project's robustness to a large degree, but the ex post scenario also proves that its internal structure is resilient even if the macro-economic environment worsens.

Table 5.14 E39 Klett-Bardshaug Road – ex-post analysis and alternative scenarios

Snapshot	Indicator									Cost		Time		Traffic		Revenue	
	FEI	InI	GI	CSI	RSI	RAI	RRI	IRA	FSI	Out-come	TIR-ESI	Out-come	TIR-ESI	Out-come	TIR-ESI	Out-come	TIR-ESI
Snapshot 1 – Award (2003)	0.738	0.840	0.563	0.722	0.200	0.333	0.667	1.000	0.719	N.A.	A⁺	N.A.	A	N.A.	A	N.A.	A⁺
Snapshot 2 – Inauguration (2005)	0.733	0.845	0.563	0.556	0.200	0.333	0.667	1.000	0.719	0	A⁺	−1	A	N.A.	A	N.A.	A⁺
Snapshot 3 – Reporting period (2014)	0.726	0.83	0.563	0.667	0.200	0.333	0.667	1.000	0.719	0	N.R.	0	N.R.	1	A	1	A⁺
Alternative scenarios																	
Ex-post consideration of low FSI for year 2003	0.738	0.840	0.563	0.722	0.200	0.333	0.667	1.000	0.506	N.A.	A⁺	N.A.	A	N.A.	A	N.A.	A⁺
Ex-post consideration of low FEI for year 2016	0.590	0.83	0.563	0.667	0.200	0.333	0.667	1.000	0.719	0	N.R.	0	N.R.	1	A_	1	A

Legend: N.A. = not available; N.R. = not relevant
Source: Authors.

5.8   A5 Maribor-Pince Motorway, Slovenia

Alenka Temeljotov Salaj, Iosif Karousos and Panayota Moraiti

5.8.1   Project delivery

5.8.1.1   Context

National Motorway Construction Programme (NMCP) has been one of the largest investment programmes in Slovenia. In 18 years, from 1990 until 2008, both the total length of motorways (from 268 to 607 kilometres) and the density of the road network (from 13.23 to 29.97 kilometres/kilometres^2) more than doubled (Carpintero, 2009). Projects, under this programme were financed through 'budget money', a term used by the Slovenian Government to describe the financing of public projects with a credit guaranteed by the State.

In the context of the NMCP, the motorway section Koper–Lendava became part of the TEN-T corridor Barcelona–Kiev, while, more specifically, the Maribor–Lendava section served the increase of traffic from Eastern-European countries (Hojs at al., 2012). The motorway also serves cities and towns in the Stajerska and Prekmurje regions (Hojs *et al.*, 2015). Notably, the existing trunk road from Maribor to Lendava functioned as a long-distance road connection to Hungary (see Figure 5.13).

Figure 5.13 A5 Maribor-Pince Motorway
Source: Bojan Salaj.

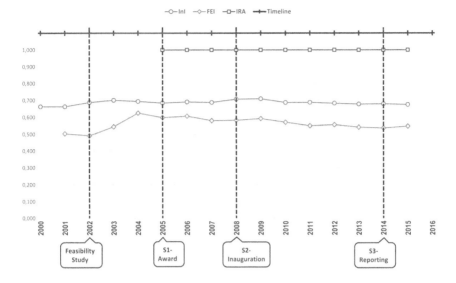

Figure 5.14 A5 Maribor-Pince project timeline

Source: Authors' compilation

The trunk road alignment is problematic, as it runs through the settlement centres resulting in poor safety and environmental conditions. The road is still of great importance for the local traffic, but is no longer used by international traffic to/from Hungary.

The motorway project relieved the settlement centres from transit traffic, while at the same time increased their tourist attraction, increased their development potential and provided better living conditions for residents in the vicinity of the motorway. The quality of the road connections was also improved, especially in terms of the accessibility to Pomurje, one of the least developed Slovenian regions.

Figure 5.14 illustrates the project's development (project timeline, IRA, FEI, and InI). The graph also depicts the project's three snapshots (S1–S4).

The A5 Maribor-Pince motorway project started in 2002 with a pre-feasibility study for the construction of road sections based on cross-section width and possible phased construction, such as basic cross-section, economical cross-section and halfway cross-section. The economical cross-section was the most favourable. In 2004, the feasibility study was prepared for the European Investment Bank (EIB). Construction started in 2005 and finished in 2008.

5.8.1.2 Contractual governance

DARS Motorway Company is a joint-stock company and the contracting authority. The company was established by law in 1993. According to the contract from 1994, the Republic of Slovenia transferred the management of all existing motorways, as well as relevant infrastructure and plant, to DARS d.d. The transfer included 198.8 kilometres of existing two- and four-lane motorways and express-ways, as well as 67.5 kilometres of respective access roads. Thus, DARS d.d. has assumed the right to collect motorway tolls as a source of income necessary for the management and maintenance of Slovenia's motorway network, as well as an important source for building new ones (source: DARS).

DARS is obliged to respect the regulations on public procurement, which stipulate that contracting authorities are obliged to publish information on public procurement procedures also on their web portals, and respecting the relevant principle of transparency of public procurement.

All phases of investment (design, construction) were initiated with open public tenders. DARS, on behalf of the Slovenian government, was the financier for all phases. The tenders for construction work were published after building permits were issued. There were 12 design contracts and tendering for construction conducted in 15 LOTS. After construction, all highways became part of the national network. DARS is responsible for operation and maintenance. All the risks were and are on the public side, as illustrated in Table 5.15.

Based on the above information, the GI was 0.563 and 0.500 after inauguration when responsibility was transferred to DARS representing the public sector (see Table 5.16).

5.8.1.3 Business model efficiency

PROJECT INTEGRATION

The A5 Maribor-Pince is well integrated into the road network. The project offers:

- efficient access to Hungarian border;
- reduction of traffic congestion in the region;
- improvement of a part of the A5 (Koper-Lendava) route and TEN-T (Barcelona-Kiev) corridor;
- efficient traffic communication nodes.

Table 5.15 A5 Maribor-Prince Motorway – risk allocation

Risks	Totally private	⟺	Totally public
Design and construction			✓
Maintenance			✓
Exploitation			✓
Commercial/revenue			✓
Financial			✓
Regulatory			✓
Force majeure			✓

Source: Authors.

DARS d.d., in accordance with the new Slovenian Motorways Company Act approved in 2010: (i) performs on behalf of and for the account of the Republic of Slovenia individual tasks regarding spatial planning, implementation of motorways and tasks related to the acquisition of real-estate needed for the building of motorways; (ii) implements on its behalf and for its account the construction of motorways; (iii) manages and maintains the motorway sections for which it acquires building concessions.

Since the very beginning (1994) of the implementation of the NMCP, 528 kilometres of four-lane, two-lane motorways and other public roads have been built and opened to traffic, which is equivalent to roads completed between 1970 and 1994. DARS has gained experience and the Maribor Pince Motorway project was well planned.

Based on the above, and in combination with the risk allocation structure, as presented in the previous section (see Table 5.16):

- the CSI was 0.245 at award (2005), decreased to 0.078 at inauguration (2008). At reporting time (2014) CSI is estimated at 0.333 reflecting the experience of the contracting authority;
- the RSI was 0.193 at award (2005), inauguration (2008) and reporting period (2014), reflecting the quite not exclusive nature of the motorway.

5.8.1.4 *Funding and financing*

FUNDING: REMUNERATION SCHEME AND PROJECT REVENUES

Operation and Maintenance is financed through direct tax sources (fuel tax, users' fees: tolls for cars and vignette trucks). Cash flow was elaborated in variants for 20- and 30-year life periods of investment. In the case of the 20-year period, cash flow was positive only in one subsection. In the case of the 30-year period, the cash flow was negative for two motorway subsections, but positive on other subsections and on the entire motorway section.

The purpose of the financial analysis was to use cash flow forecasts for the calculation of the financial internal rate of return of investment. The latter was calculated regardless of financial sources and showed the ability of covering investment costs (construction costs, toll collecting costs and maintenance costs) with revenues from the operation of the road (toll revenues).

The motorway is open to all private vehicles and freight traffic. A5 Maribor-Pince Motorway serves, however, vehicles for commuting and logistics purposes, accessing different travel destinations and roads along the A5 national road or the TEN-T international corridor.

Based on the above information (see Table 5.16):

- the RAI was initially 0.667 as low risk direct taxation was considered and then improved to 1.000 at reporting as scheme proved to be very low risk;
- the RRI was 1.000 throughout the project lifetime.

FINANCING SCHEME

The project was financed through the public budget and, also, included loans from theEIB. The estimated value of the construction was 628 M EUR.

Based on the above, the FSI was 0.970 throughout the project lifetime, considering a 20 per cent contribution from the EIB.

5.8.1.5 Outcomes

The project met its targets with respect to all outcomes.

The main success was a new highway for local residents and other users. The old road is no longer congested and travel time is reduced.

COST AND TIME TO CONSTRUCTION COMPLETION

The A5 Maribor-Pince Motorway was built on time and within budget.

- The cost of investment was *almost* 10 per cent lower than was expected, due to better geo-mechanical conditions (less construction work). The impact of the agreed request of complementary works on cost was around 5 per cent – some claims concerned unit costs.
- There were no delays in the initiation of works following award and no delays in the completion of works.

ACTUAL VERSUS FORECAST TRAFFIC

Actual traffic is higher compared to forecasts. The motorway was fully operational in 2008. In 2004, the measured level of traffic on the existing G1 road shows an average daily traffic of 14,500 vehicles from Maribor to Lenart, including 20 per cent of trucks, compared to 5,600–8,700 vehicles per day in 2000. A5 Maribor-Pince motorway is the most loaded motorway in Slovenia in terms of heavy truck traffic. Data in 2014 shows that 18 per cent of the 23,226 vehicles average daily traffic is heavy trucks (more than 7t), namely 4,043 trucks (source: DARS).

ACTUAL VERSUS FORECAST REVENUES

Actual revenues were higher than forecasted, due to several HGVs travelling from East to West and vice versa.

PROJECT RATING OVER TIME

Table 5.16 presents the results of the case snapshot analysis for the A5 Maribor-Pince project, based on the indicator values, the project outcomes and the TIRESI ratings over its lifetime.

The project outcomes in terms of revenue levels match the related high TIRESI rating scores. Other TIRESI ratings are more conservative considering the implementation context.

Table 5.16 A5 Maribor-Pince Motorway – ex-post analysis and alternative scenarios

Snapshot	Indicator									Cost		Time		Traffic		Revenue	
	FEI	Inl	GI	CSI	RSI	RAI	RRI	IRA	FSI	Out-come	TIR-ESI	Out-come	TIR-ESI	Out-come	TIR-ESI	Out-come	TIR-ESI
Snapshot 1 – Award (2005)	0.598	0.685	0.563	0.245	0.193	0.667	1.000	1.000	0.970	N.A.	B_{EX}	N.A.	B_{EX-}	N.A.	B_{EX}^+	N.A.	A_-
Snapshot 2 – Inauguration (2008)	0.583	0.707	0.563	0.078	0.193	0.667	1.000	1.000	0.970	1	N.R.	1	N.R.	N.A.	B_{EX}^+	N.A.	A_-
Snapshot 3 – Reporting period (2014)	0.536	0.679	0.500	0.333	0.193	1.000	1.000	1.000	1.000	1	N.R.	1	N.R.	1	A	1	A^+
Alternative scenarios																	
Ex-post consideration of PPP model for year 2005	0.598	0.685	0.688	0.514	0.214	1.000	1.000	1.000	0.620	N.A.	B_{EX}^+	N.A.	B_{EX-}	N.A.	A	N.A.	A^+

Legend: N.A. = not available; N.R. = not relevant
Source: Authors.

5.8.2 *Alternative scenarios of project delivery*

In its actual public delivery form, no significant improvements can be achieved with respect to the project's structural indicators (CSI, RSI, GI) following the award snapshot. In fact, in the last snapshot (2014), these indicators achieve their threshold scores.

Therefore, and given the fact that the PPP delivery model has never been adopted in Slovenia, this alternative scenario of project delivery is examined. Accordingly, a PPP scenario is tested whose financing scheme includes a combination of private equity capital together with governmental subsidies, while the project's funding scheme remains the same. In this case, the FSI value would decrease to approximately 0.620.

As shown in Table 5.16, this first alternative scenario of project delivery yields improved TIRESI rating for all outcomes (except time). The choice of a PPP delivery scheme may thus have been a viable choice.

5.8.3 *Concluding remarks*

The A5 Maribor-Pince Motorway case study is a publicly financed project with a user charges based funding scheme.

The rating of cost to completion, time to completion and traffic outcomes were rated more conservatively due to the prevailing implementation context. The potential likelihood of the revenue outcome is accurately represented by the TIRESI rating in most case snapshots.

The ex-post analysis showed that in a country with no PPPs, the project could have been procured as a PPP with a likelihood of achieving better outcomes.

5.9 Athens Ring Road (Attica Tollway), Greece

Iosif Karousos and Panayota Moraiti

5.9.1 *Project delivery*

5.9.1.1 *Context*

Attica Tollway is a pioneering project constructed on a concession basis, constituting one of the biggest (privately) (co)financed road projects in Europe. It belongs to the first generation of co-financed projects awarded in Greece during the 1990s, essentially paving the way and laying the foundations for the execution of future successful concession contracts, both in Greece and in other European countries. Attica Tollway is an urban motorway and part of the TEN-T, connecting the 30 municipalities of the Attica basin, allowing quicker access to areas, which, before its construction, were highly congested and required a great amount of travel time.

During its construction and operation, the Attica Tollway introduced a number of innovations to the Greek construction and motorway operation sector (Halkias *et al.*, 2013). It was also constructed in parallel with flood protection works (contract value of EUR 791 M), as the Attica Tollway passes through the three large hydrographic basins of Attica, rendering the construction of substantial extensive flood protection works within the scope of the project implementation imperative. The flood protection works constructed were dimensioned to be adequate for the existing and future land use.

The idea of building the Attica Tollway dates back to 1963, and was included in the first ever regional traffic planning study for the city of Athens and its metropolitan area. Sprawling development to the north of Athens over the years, the decision in the late 1970s to build the new airport in its present location at Mesogeia, and the decision to build a city connector road along the foothills of the Mountain of Imittos in the early 1990s, resulted in the departure from the concept of a 'ring' road, and transformed the Attica Tollway into an urban tollway as shown in Figure 5.15.

The project's real advancement began in 1985, when it became part of the official transport infrastructure plans for metropolitan Athens, along with the (unsuccessful) bid to host the Centennial Olympic Games in 1996. It was then in the early 1990s that the Greek Ministry of Public Works adopted the method of co-financing for the road through a Build–Operate–Transfer contract.

Technical characteristics:
Total length: 66.7 km with 3 traffic lanes in each direction and emergency lane, separated by a traffic island reserved for the operation of the suburban railway.

Other features:
Other main lines: 31.33 km; Network Utility / side roads: 150 km; Interchanges: 32; Road bridges (Overpasses): 104; Road bridges (underpasses): 38; Rail Bridges: 37; Footbridges: 15; Tunnels (Cut & Cover): 63; Total length of tunnels and Cut & Cover: 15.64 km; Anti-flood works: 66.7 km; Total area of support facilities premises: 122,000 sq.m.

Figure 5.15 Attica Tollway
Adapted source: Attiki Odos S.A.

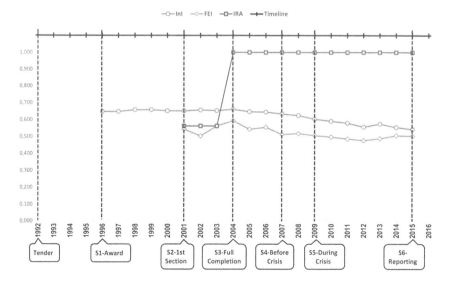

Figure 5.16 Attica Tollway project timeline
Source: Authors' compilation.

Construction works started in 1997 and the motorway was given to traffic in phases/sections. The first one opened to traffic in March 2001, achieving the milestone of serving the newly constructed Athens International Airport. The last section was opened to traffic in 2004.

Figure 5.16 illustrates the project's development (project timeline, IRA, FEI, and InI). The graph also depicts the project's six snapshots (S1–S6).

5.9.1.2 Contractual governance

An international tender was announced in 1992 and in March 1996, the project was awarded to the lowest bidder of the three international consortia that participated in the process. The Ministry of Environment, Physical Planning and Public Works was responsible for the preparation of the tender documents and the tendering procedure. The Concession Contract was ratified by the Greek Parliament as an Act of law on the 23 May 1996.

The Concession Agreement includes a safety mechanism, securing the interests of the Greek State through a maximum Return on Equity. The Concession period will extend for a maximum of 25 years (including construction period), or will end earlier, in the case that the maximum Return on Equity (11.6 per cent) is reached.

Expropriation cost was undertaken entirely by the Greek State. Loan guarantees were provided by the concessioner during construction and by the State for the operations phase. The environmental risk was undertaken by the State in reference to environmental law changes. The Greek State has proceeded to design changes in the Hymittos Western Peripheral Section for environmental reasons, after the concession commencement. Variation orders were also related to water management in the area of the project. On the other hand, the compliance with environmental guidelines was solely attributed to the SPV.

A project of such scale met significant difficulties during its realisation. The financial close was delayed, mainly because of uncertainties surrounding the project. These uncertainties increased the risks for the banks, delaying the signing of the financial agreement and forcing the public sector to provide funds to the sponsors to begin construction before financial close was reached. Other difficulties were due to variation orders issued by the State, mainly for environmental reasons, which involved significant design changes.

Furthermore, the concession contract did not include mechanisms for extensions of time and delay make-up in case of State-instructed variations. Solutions were found following negotiations between all parties involved, whereby amendments to the concession contract were introduced, leading to the satisfaction of the banks and the reach of financial close. Risk allocation is depicted in Table 5.17.

Based on the above information, the GI was 0.688 throughout the project lifetime.

Table 5.17 Attica Tollway – risk allocation

Risks	⟺		
	Totally private		*Totally public*
Design and construction	✓		
Maintenance	✓		
Exploitation	✓		
Commercial/revenue	✓		
Financial		✓	
Regulatory			✓
Force majeure			✓
Environmental			✓

Source: Authors.

5.9.1.3 Business model efficiency

PROJECT INTEGRATION

The Attica Tollway is a unique project offering:

- the only ring road for the Metropolitan area of Athens, which integrates the full road network for fast and safe transport in the entire Attica region and contributes to the integrated regional/urban plan for Attica;
- the only motorway connection to Athens International Airport;
- reduction in traffic congestion in the main arteries of Athens;
- a crucial segment of the PATHE TEN-T (priority axis 7) motorway by connecting the two main National Roads of Greece (National Road of Patras-Athens and National Road of Athens-Thessaloniki);
- direct connection between the western and the eastern areas of Attica.

COMPETENCE OF THE CONTRACTING AUTHORITY

The Attica Tollway project was planned on a central government level, by the Ministry of Development, Competitiveness, Infrastructure and Transport Networks (previously called Ministry of Environment, Physical Planning and Public Works). EYDE/LSEP is the special agency of the Ministry, which undertakes the supervision of the motorway's operation and maintenance. Central government was directly involved in all stages of the development, design, tendering and negotiation procedure, while it is also overseeing operation and maintenance.

The project was well-planned including life cycle planning, as additional objectives were:

- to access a future spur (presently served by a four to six lane arterial road) that will connect to the passenger and cargo harbour of Rafina, which offers quicker connections to central and northern Aegean islands than the main port of Piraeus;
- to support significantly the flood prevention system for the entire region of Attica by substantial interventions in the three main hydrographical basins of Attica, that is the Thriasio Plain, the Athens Basin and Mesogeia.

The Contracting Authority successfully managed the issue of tolls being imposed on an urban axis, which was a first not just for the Athens metropolitan area but for Greece as a whole. In effect, in the very early stages of the project, there was public opposition with respect to the flat rate toll imposed as opposed to toll rates based on distance travelled. However, this opposition was soon overcome, especially as it became clear that the scope of the infrastructure was to serve traffic traversing the metropolitan area and not traffic between suburbs. However, it should be noted that this was one of the first PPP projects in Greece and the Contracting Authority had, by default, limited experience.

COMPETENCE OF THE PRIVATE SECTOR ACTOR(S)

Attica Tollway S.A. was formed as a joint venture of almost all large Greek construction companies (AKTOR, AVAX, ALTE, ATTI-KAT, HELLENIC TECHNODOMIKI, ETETH, SARANTOPOULOS, PANTECHNIKI, TEV, TEG, EGIS PROJECTS). 'ATTIKI ODOS S.A.', is the Concession Company (Concessioner) of the project, which has undertaken the design, construction, financing, operation and maintenance of the motorway, through the execution of a Concession Contract with the Greek State. The Concessioner has established Contracts back to back with the Concession Agreement with 'ATTIKI ODOS CONSTRUCTION JOINT VENTURE' for the project construction and with 'ATTIKES DIADROMES S.A.' (also known as Attica Tollway Operations Authority) for the operation and maintenance of the project.

A subsequent consolidation in the Greek construction industry has led, through bankruptcies, mergers and acquisitions to the formation of a few large construction groups. The Attica Tollway consortium was a catalyst for this consolidation.

The Concession Company's shareholders structure currently (31 December 2013) consists of:

- AKTOR CONCESSIONS S.A. (member ELLAKTOR S.A. Group) 59.25 per cent;
- J.&P. AVAX S.A. 21.00 per cent;
- ETETH S.A. (member of J.&P. AVAX S.A. Group) 9.82 per cent;
- PIREAUS-ATE BANK S.A. 9.88 per cent and
- Transroute International 0.04 per cent.

Based on the above, and in combination with the risk allocation structure as presented in the previous section (see Table 5.18):

- the CSI was 0.083 at award (1996), increased to 0.566 (2001 and 2004) and dropped after the economic crisis to 0.489 (2007, 2009 and 2015);
- the RSI was 0.224 at award (1996) and inauguration of first section (2001), increased slightly to 0.229 at full project completion (2004) and thereafter (2007, 2009), recently dropping to the initial value of 0.224 (2015).

5.9.1.4 *Funding and financing*

FUNDING: REMUNERATION SCHEME AND PROJECT REVENUES

The motorway is *open* to all private vehicles and freight traffic. Attica Tollway serves, however, mainly passenger cars and the main purpose of its use is commuter travel. In addition, many large logistics centres have emerged or relocated to the western part of the Tollway, since this location combines large open spaces and quick access to ports, the railway and National Roads.

Apart from minor revenues connected to real estate, the remuneration and revenue scheme is based on real tolls. The contract provides a maximum toll rate that can be charged. Revenues were expected to cover costs and obligations.

Based on the above information (see Table 5.18):

- the RAI was 0.333 from award (1996) to the inauguration of first section (2001), increased to 0.667 at full project completion (2004) until before the economic crisis (2007), and then dropped again to 0.333 since the economic crisis (2009, 2015). Notably, at initiation the risk related to tolls was high. As user acceptability of the toll scheme increased, risk related to toll charging could have been reduced in the estimation of the RAI. However, this was just before the crisis and, therefore, high risk continues to exist;
- the RRI was 0.667 from award (1996) to the inauguration of first section (2001), increased to 0.833 at full project completion (2004) until before the economic crisis (2007), and then dropped again to 0.667 since the economic crisis (2009, 2015).

FINANCING SCHEME

The financing of the project has been ensured through State contributions including European Commission (EC) Structural Cohesion Funds, private equity and loans. Commercial banks involved include the following: Bank of Tokyo-Mitsubishi, HypoVereinsbank, Commercial Bank of Greece, HSBC Athens, National Bank of Greece, Société Generale, European Investment Fund, ABN AMRO Bank NV, Agricultural Bank of Greece, Alpha Credit Bank, Banca Monte dei Paschi di Siena (London), Bank of Scotland, De Nationale Investeringsbank NV, Piraeus Bank Greece, EIB, ING Bank NV and Ergobank.

The construction was financed (total 1,346,309,241 EUR) as follows:

- equity: 157.6 M EUR
- loans: 666.2 M EUR (the majority were EIB loans)
- Greek State: 420.6 M EUR
- income from interest: 12 M EUR
- income from operation (during construction): 90 M EUR

Notably, the project reached financial close in March 2000. Until then, the project was financed by the Greek Government.

Based on the above, the FSI was (see Table 5.18) 1.000 at award (1996), dropping then to 0.561 following financial close (2001, 2004, 2007, 2009, 2015) when private sources of financing were included into the financing scheme.

5.9.1.5 Outcomes

Attica Tollway is a successful project and has exceeded forecast demand. The reason is that Attica Tollway has produced significant improvements to traffic conditions in the metropolitan area, as well as benefits in the economy and overall metropolitan infrastructure. In addition, it has achieved enormous public acceptance despite the initial resistance and the imposition of an open and flat toll regime.

On the other hand, the global financial crisis could have affected the project significantly. Between 2009 and 2013, the country's GDP had fallen by 20 per cent, while unemployment had risen from 8 to 29 per cent. Most of this negative impact is concentrated in the Athens metropolitan area where approximately 50 per cent of the country's population is located. Taking into account the drop in car ownership and the increase of fuel prices and taxation, it is considered that the project has shown significant resilience and is still viable.

The motorway has a very high rate of acceptability. The company runs biannual user satisfaction surveys that indicate 90 per cent user satisfaction.

Regarding project goals, current studies show that the average vehicle-trip time is reduced by 28 minutes when choosing the Attica Tollway. There is considerable fuel consumption reduction, while reliability was improved fully surpassing initial expectations (ex post – observed – share of delayed traffic: less than 1 per cent). The same applies to safety with ex post (observed) total accidents per 1000 vehicle-kilometre by year equal to 0.000547.

COST AND TIME TO CONSTRUCTION COMPLETION

Attica Tollway was built on time and on budget and was the backbone of the transportation network during the execution of the Athens Olympic Games in 2004. The problems faced during construction were solved owing to the good faith negotiations held between all parties involved.

ACTUAL VERSUS FORECAST TRAFFIC

The forecast AADT was estimated to level off at approximately 245,000 vehicles after 10 years of operation with a gradual increase from 160,000 in 2004.

The motorway was fully operational as of mid-2004 and presented AADT 30 per cent above forecast in the years 2004–2009 (peak traffic in 2009 307,993 AADT). Ever since, and due to the economic crisis, AADT has dropped to 200,449 in 2013 (281,217 in 2010, 250,625 in 2011 and 215,767 in 2012). AADT has since 2013 levelled off at approximately 200,000.

ACTUAL VERSUS FORECAST REVENUES

From the start of operation and until 2011, the actual traffic was higher than the predicted reflecting on the respective revenue stream. However, for the first time in 2012, this changed and the actual traffic was lower than the predicted one. Traffic levels have remained stable since 2013. Based on annual financial reports, Attica Tollway is a profitable project. The report for year 2013 showed net profits of 68.7 M EUR (before tax) and 20.9 M EUR (after tax). Net profits dropped by 19.2 M EUR (net after tax) in relation to 2012. As the traffic drop has stabilised, it is considered that this level of profitability will be sustained. No changes have been made to toll prices since 2010.

PROJECT RATING OVER TIME

Table 5.18 presents the results of the ex-post analysis for the Attica Tollway project, based on the indicator values, the project outcomes and the TIRESI ratings over time. It is evident that the implementation context of the project affects the potential of success.

5.9.2 *Alternative scenarios of project delivery*

As indicated through the review of Table 5.18, the key factors limiting the potential to reach targets have been the exogenous ones, expressed through the implementation context indicators (i.e. the FEI and the InI). These limitations reflect the project's vulnerability to exogenous factors. Hence, the potential of enhancement is focused on the endogenous factors.

Following award and considering the nature of the project (road infrastructure), little improvement may be achieved with respect to the CSI and RSI, while the GI has already achieved its threshold score (GI = 0.500).

Another potential measure would be the increase of the value of the FSI rendering it greater than 0.600. This would require low cost financing, which is deemed unrealistic, especially as the project reached financial close two years into its construction. The ex post approach considered herewith is one that has been only recently employed: crowdfunding.

Today, equity crowdfunding is increasingly offering an alternative or a complementary type of financing by engaging local constituents in infrastructure development. It is based on the motivation of local residents (and others) that find benefit in the realisation of the project. Notably, such anticipation of benefits existed at the early stages of the project. Figure 5.17 depicts the annual average index of price of dwellings in Athens in the period 1993–2015. A continuous increase in values is recorded from 1993, three years prior to the project award up until the beginning of the financial crisis in 2009. Given that the project has a significant economic impact in terms of supporting development in a region that was previously underdeveloped and rural, local residents would not only benefit from returns on the project as financiers via crowdfunding, but also largely from the increase of their individual land value. Obviously, the latter represents the motivation.

Based on the above, a 10 per cent equity capital from crowdfunding is assumed in the project financing structure with an associated cost of capital of 7 per cent. This contribution results in a minor increase of the FSI from 0.561 to 0.606. This would retain the positive rating and outcomes of the project at a lesser cost of capital.

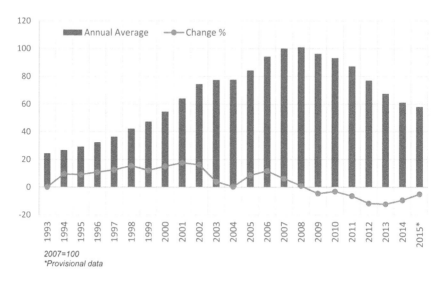

2007=100
*Provisional data

Figure 5.17 Index of prices of dwellings in Athens (1993–2015)

Sources: Author's compilation based on Bank of Greece data collected by the credit institutions (2006 onwards, apartments only) and calculations based on "Property Ltd" (1997–2005, all dwellings) and "Danos and Associates" (1993–1997, all dwellings).

Table 5.18 Attica Tollway – ex-post analysis and alternative scenarios

Snapshot	Indicator									Cost		Time		Traffic		Revenue	
	FEI	InI	GI	CSI	RSI	RAI	RRI	IRA	FSI	Out-come	TIR-ESI	Out-come	TIR-ESI	Out-come	TIR-ESI	Out-come	TIR-ESI
Snapshot 1 – Award (1996)	0.543	0.648	0.688	0.083	0.224	0.333	0.667	0.250	1.000	N.A.	B_{EX-}	N.A.	B_{EX}	N.A.	B_{EX-}	N.A.	B_{EX-}
Snapshot 2 – Inauguration of 1st section (2001)	0.543	0.652	0.688	0.566	0.224	0.333	0.667	0.563	0.561	N.A.	B_{EX}^{+}	N.A.	A_{-}	N.A.	A_{-}	N.A.	A
Snapshot 3 – Project completion (2004)	0.593	0.661	0.688	0.566	0.229	0.667	0.833	1.000	0.561	0	B_{EX}^{+}	0	B_{EX-}	1	A	1	A^{+}
Snapshot 4 – Before crisis (2007)	0.513	0.634	0.688	0.489	0.229	0.667	0.833	1.000	0.561	0	N.R.	0	N.R.	1	A_{-}	1	A
Snapshot 5 – During crisis (2009)	0.507	0.603	0.688	0.489	0.229	0.333	0.667	1.000	0.561	0	N.R.	0	N.R.	–1	B_{EX}^{+}	0	B_{EX}^{+}
Snapshot 6 – Reporting period (2015)	0.504	0.543	0.688	0.489	0.224	0.333	0.667	1.000	0.561	0	N.R.	0	N.R.	–1	C+	0	B_{EX}^{+}
Alternative scenarios																	
Ex-post consideration of crowd funding in year 2001	0.543	0.652	0.688	0.566	0.224	0.333	0.667	0.563	0.606	N.A.	B_{EX}^{+}	N.A.	A_{-}	0	A_{-}	0	A

Legend: N.A. = not available; N.R. = not relevant
Source: Authors.

5.9.3 Concluding remarks

The Atticka Tollway has been a project with marginally resilient internal structure, which was from its planning phase vulnerable to the external context. The decision to structure it as a PPP with the remuneration scheme relying entirely on user tolls added to the project's vulnerability. Nevertheless, the project did prove both successful and resilient surpassing the likelihood suggested by its TIRESI ratings.

One of the crucial points in the project's life cycle was its financing. The financial close of the project was reached a few years into its construction and just before its first section was opened to traffic. The high cost of financing involved, which also pushed for a larger contribution from the public sector, tested the project's financial viability in the early years.

As an ex post enhancement scenario, crowdfunding was selected to contribute 10 per cent to project financing. Today, it is assumed that local residents, stemming from the increase of real estate values that were prompt from the project initiation (tendering procedures), would have been motivated to contribute as financiers to the project. Had this scenario been realised, the project's rating would have remained unchanged but the project would have been implemented at a lower cost of capital with lesser government contribution. Such a scenario was not possible or considered at the time.

5.10 Port of Agaete, Canary Islands, Spain

Javier Campos, María Manuela González-Serrano, Lourdes Trujillo and Federico Inchausti-Sintes

5.10.1 Project delivery

5.10.1.1 Context

Agaete is a small village (pop. 5600 in 2014) located in the northwest corner of the island of Gran Canaria, in the Canary Islands. Until the beginning of the 1980s, the local economy was relatively isolated and centred on agriculture, although the village had a small fishing harbour with minimal facilities for local fishermen. During the 1990s, the 30-kilometre road that connected Agaete with the rest of the island and the capital city, Las Palmas de Gran Canaria (pop. 650,000 in 2014, including its surrounding metropolitan area), was greatly improved, and the village became a crucial maritime link within a regional government-promoted transport network to connect Gran Canaria with Tenerife, the second main island in the archipelago.

Since the end of the 1970s, local authorities had already claimed for improvements in their small port. A first project was designed by the Spanish government in 1981 through the Ministry of Public Works in Madrid. The project was not well received by local residents because it was based on a large dock, whose huge seawall had a high visual impact. Despite opposing public opinion, the works were awarded in 1982 in a competitive tendering process to SATO (Sociedad Anónima de Trabajos y Obras), a Spanish building company with solid national experience. It was the only company that participated in the tender.

However, both political and social pressure increased, and works were halted just one year after their commencement in search for an alternative design. The modified project included a new dock with a special design to alleviate the impact of sea waves during bad weather conditions, which was also heavily criticised by local authorities and had to be once again elaborated.

In 1985, the administration and management of local ports were transferred from the central government to the regional government (Gobierno de Canarias), which updated the existing project allowing for public works to restart by the same construction company re-founded as SATOCAN with local capital. In 1987, an agreement between the regional government and the island government (Cabildo Insular) was reached to update the budget and speed up the works. Works were finally concluded in 1993, although the commercial use of the port was negotiated between the regional and central government resulting to additional improvements to facilitate its usage as such, being implemented in later years.

The port of Agaete is essentially a monopoly, with the only competitor for the Gran Canaria-Tenerife Ro-Ro traffic being the Port of Las Palmas at a distance of 32 kilometres. The commercial use of the Port of Agaete was awarded as a monopoly to a single operator, Fred Olsen, in 1994, for safety reasons. Nonetheless, the advent of this single new operator broke the former monopoly

Figure 5.18 The Port of Agaete
Source: www.surcando.com

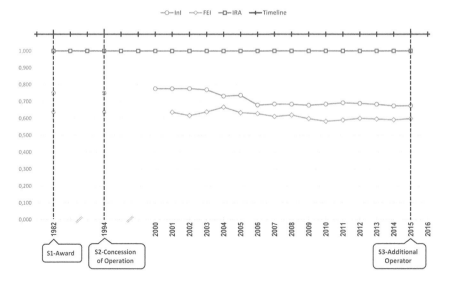

Figure 5.19 Port of Agaete project timeline
Source: Authors.

and forced a decrease in prices. In 2015, passenger services opened to competition and a second license was awarded to a competitor operator (Líneas Armas)[1].

Finally, the global economic crisis that hit the Spanish economy since 2007 has not affected the traffic demand of the port. Both traffic and revenues remain high in spite of the crisis.

Figure 5.19 illustrates the project's development (project timeline, IRA, FEI, InI). The graph also depicts the project's three snapshots (S1–S3): 1982, year of award; 1994, the year the first shipping company (Fred Olsen) began operations and 2015, when the additional operator (Naviera Armas) came into play.

5.10.1.2 Contractual governance

The Spanish Ministry for Public Works, through the Administrative Ports Planning Group (Grupo Administrativo de Puertos) contracted the initial works. After 1986, the works and supervising competencies were transferred to the regional government (Gobierno de Canarias), first through the Dirección General de Puertos (an administrative body), and, since 2012, through Puertos Canarios, a public agency.

The initial works were carried out under Spanish legal framework for ports and public works (more specifically, the Law for Public Contracts). In 1985, the RD2250/85 (government decree) transferred the management of the small and medium-sized ports to the Government of the Canary Islands (a Spanish Autonomous Community since 1982). The use of Agaete as a commercial port was negotiated in 1992 and the Regional Statute was suitably adapted in 1996 (Ley

Orgánica 4/96). Today, the legislative framework is defined in the Law of Ports of the Canary Islands (Ley 14/2003).

As described in the above, national authorities initiated the project, but competencies were later transferred to the regional government. The local government (the island's government also participated in the final financing) was involved at the beginning of the project and in 1996, to authorise the commercial usage. Since then, all the responsibility is assumed by the regional government.

Risk allocation is summarised in Table 5.19. Risks are mainly borne by the public party. The infrastructure and the port belong to the public authority that negotiates its exploitation with the concessioner. In this sense, Fred Olsen pays for the usage of the facilities and the warehouses. The commercial exploitation (passenger and cargo transport) belongs entirely to the private party.

Both the construction and operation contracts were fairly standard. They were designed by the Spanish Government and included clauses and provisions that were common in the Spanish legislation, including penalties for unjustified delays (in the construction stage). The successive changes in the project did not imply changes in the contractor, but new contracts had to be signed to update the clauses.

Based on the above, the GI is 0.500 throughout the entire lifetime of the project.

5.10.1.3 Business model efficiency

PROJECT INTEGRATION

Agaete has become the crucial maritime link between Gran Canaria and Tenerife. In the 1990–2000s, it was the centre of an ambitious regional plan to create 'a motorway' between the two main islands, with improvements in the connecting roads and access from Las Palmas de Gran Canaria through a motorway. There is also no integration between the port of Agaete and other commercial ports.

Table 5.19 Port of Agaete – risk allocation

Risks			⟺		
Design				✓	
Construction			✓		
Maintenance				✓	
Exploitation	*Totally private*	✓			*Totally public*
Commercial/revenue			✓		
Financial		✓			
Regulatory				✓	
Force majeure					

Source: Authors.

The shipping company Fred Olsen offers tourist packages for visiting a number of attractions on both islands. It also offers a (free) bus shuttle service from Las Palmas de Gran Canaria to Agaete and from Santa Cruz de Tenerife to Los Cristianos in the South of the islands, where the company operates three routes to El Hierro, La Gomera and La Palma. This service eases the connection by road between the city and Agaete, while also strengthening its demand compared to the airport.

COMPETENCE OF THE PRIVATE SECTOR ACTOR(S)

The port construction and refurbishment project was awarded to SATO (later re-founded as SATOCAN) after a competitive tendering that did not attract many participants. In fact, SATO was the only final bidder.

The operation of ferry transport services was directly awarded to *Fred Olsen* that was the only company interested. Apparently, another company (*Trasmediterranea*) also expressed its initial interest, but was discouraged due to technical reasons (not enough capacity at the dock to operate with safety). Both SATOCAN and Fred Olsen are well-known companies in their respective sectors. SATOCAN have worked at a national level, while Fred Olsen is a Norwegian company with international experience.

Based on the above and in combination with the risk allocation structure as presented in the previous section (see Table 5.20):

- the CSI was -0.100 at award (1982), increasing to 0.000 at the concession of operation (1994), finally reaching 0.296 in 2015;
- the RSI remained stable at 0.223 throughout the project lifetime.

5.10.1.4 Funding and financing

FUNDING: REMUNERATION SCHEME AND PROJECT REVENUES

Today, the main usage of the port is the provision of regular passenger and Ro-Ro services, while project revenues constitute user charges. The main user of the Port of Agaete is Fred Olsen, the private-owned shipping company that provides regular ferry services between the two main islands (Gran Canaria and Tenerife). Local fishermen and private recreational yachts also make occasional use of the safe dock.

Concessioners pay a fee to the government. In 2012, the regional government obtained 1.2 M EUR from public prices for the usage of the dock facilities (by Fred Olsen). It also obtained 17 M EUR for renting warehouses and terminal facilities.

Based on the above, the RAI and the RRI are both 1.000 for all snapshots (see Table 5.20). Notably, remuneration is based on user charges but they have been deemed low risk.

FINANCING SCHEME

The project was contracted in accordance to the Spanish Law for public contracts through an open procedure. It was divided in several stages, at the end of each one the builder received the corresponding payment (standard building contract). All the payments were made by the Spanish Government after the corresponding budget assignments.

There is a lack of suitable financial information for this case. The FSI is 1.000 for all three snapshots.

5.10.1.5 *Outcomes*

The port of Agaete can be seen as an example of public intervention that radically transformed the socioeconomic context of a small fishing village into an important maritime hub in the Canary Islands. The new infrastructure brought new services between Gran Canaria and Tenerife and substantially improved the performance of the previous port. Had the project not been carried out, the isolation of the village would have possibly continued for several years.

The port of Agaete is considered a successful project having changed dramatically the lives and prospects of the inhabitants of the northwest corner of Gran Canaria by creating a relevant link for inter-island traffic. Beyond its economic impact, the port is now a vital entry for the island and a way to reduce the political differences among islands. The level of satisfaction is high.

Until the 1980s, maritime travel time between Gran Canaria and Tenerife for passengers and services was 4–4:30 hours (from the port of Las Palmas de Gran Canaria) and there was only one (public) operator, TRASMEDITERRANEA, whose services lacked reliability. The main competitive advantage of the new service (from Agaete to Tenerife) was that the time on-board was reduced to one hour.

The key of the success of the port of Agaete lies in the capacity of the authorities to identify the lack of a closer and faster connection by sea between the two main islands of The Canary Islands. Agaete has the port closer to Santa Cruz of Tenerife. Moreover, the shipping company (Fred Olsen) operates with fast catamarans (Ro-Ro services). Both the shorter distance and the faster vessels reduce the travel time by one and a half hour compared to the aforementioned services from city to city. Finally, the advent of the new competitor broke the former monopoly producing a reduction in prices in the service.

COST AND TIME TO CONSTRUCTION COMPLETION

The project suffered from both cost and time overruns. The initial budget was 3.1 M EUR in 1982, but was updated in 1987 to an additional 2.6 M EUR amounting to a total of 5.7 M EUR. In terms of time overrun, the delay was caused by governmental issues.

Table 5.20 Port of Agaete – ex-post analysis

Snapshot	Indicator									Cost		Time		Traffic		Revenue	
	FEI	Inl	GI	CSI	RSI	RAI	RRI	IRA	FSI	Out-come	TIR-ESI	Out-come	TIR-ESI	Out-come	TIR-ESI	Out-come	TIR-ESI
Snapshot 1 – Award (1982)	0.637	0.700	0.500	–0.100	0.223	1.000	1.000	1.000	1.000	N.A.	B_{EN}	N.A.	C^+	N.A.	A	N.A.	A^+
Snapshot 2 – Concession of operation (1994)	0.637	0.749	0.500	0.000	0.223	1.000	1.000	1.000	1.000	0	N.R.	0	N.R.	1	A	1	A^+
Snapshot 3 – Introduction of additional operator (2015)	0.598	0.674	0.500	0.296	0.223	1.000	1.000	1.000	1.000	–1	N.R.	–1	N.R.	1	A	1	A^+

Legend: N.A. = not available; N.R. = not relevant

Source: Authors.

ACTUAL VERSUS FORECAST TRAFFIC

The port of Agaete is a success regarding traffic expectations, especially because it led to an increased mobility of residents and tourists. According to the Statistics Institute of the Canary Islands (ISTAC), the rate of growth of passengers and vehicles was a 4 and 7 per cent, on average, from 2011 to 2016, respectively. The global economic crisis that hit the Spanish economy since 2007 has not affected the high demand. Both traffic and revenues remain high in spite of the crisis.

ACTUAL VERSUS FORECAST REVENUES

The revenues have been practically 'ensured' from the very beginning of the project because of the high demand. The company is seeing a considerable increase in passengers since 2000.

5.10.2 Conclusions and project rating over time

The TIRESI in its current development does not provide ratings for port projects, especially concerning commercial ports.

Demonstrating the TIRESI rating potential, the key characteristics of the port of Agaete were considered. In principle, its function as a link connecting to isolated locations was considered and the project was assessed according to the bridge and tunnel rating assessment.

Based on this assumption all indicators were calculated as described in this case description.

The ratings over time are presented in Table 5.20, also including the indicator values over time.

Table 5.20 presents the results of the ex-post analysis for the Port of Agaete project, based on the indicator values, the project outcomes and the BENEFIT ratings over time. The likelihood to achieve respective outcomes corresponds with the observed project outcomes and confirms the approach.

5.11 Conclusions

Nine case studies of transport infrastructure projects from nine different countries and dealing with different transport infrastructure modes have been described in this chapter. Each case study has two parts: the first describes an ex-post analysis of the actual project delivery in terms of the implementation context, contractual governance, business model efficiency, funding and financing and project outcomes. The second part investigates actions or measures that could have improved the outcomes of the project, 'what if' scenarios, while two cases test the limits and applicability of the TIRESI rating system. A summary of the measures proposed in the nine case studies is provided in Table 5.21 along with the description of the respective implementation context (FEI and InI). These nine cases demonstrate some of the potential applications of the TIRESI rating.

Table 5.21 Alternative scenarios and enhancement measures tested

| Project Case | Scenarios | Exogenous indicators | | Endogenous indicators | | | Policy indicators | | |
| | | Implementation context | | Structural indicators | | | | | |
		FEI	InI	GI	CSI	RSI	RAI	RRI	FSI
Combiplan	Enhancement testing for rail	>0.70	>0.80	✗	✗				
Horgos – Novi Sad	Enhancement	<0.50	<0.50	✗	✗				
Blanka Tunnel	Enhancement	>0.60	>0.65	✗	✗				
Reims	Enhancement	>0.65	>0.70	✗	✗	✗	✗		
Brabo 1	Testing the fuzziness of the threshold values	>0.65	>0.75	≈0.70					
E39 Klett – Barshaug	What if FEI <0.59? Enhancement	>0.70	>0.83						0.506
A5 Maribor-Pince	What if FSI=0.62?	[0.54, 0.60]	>0.65						0.62
Attica Tollway	What if crowdfunding 10%?	<0.60	[0.60, 0.65]						0.606
Agaete Port	Testing for seaport link	>0.60	>0.65						

Source: Authors.

5.11.1 Background conditions and actionable variables

In the BENEFIT framework, indicators are distinguished as those describing background conditions and actionable variables. The exogenous indicators FEI and InI reflect the background conditions and are non-actionable because they cannot be influenced by the decision makers involved in a project. The endogenous (structural and policy) indicators provide varying levels of flexibility to decision makers. The structural indicators CSI, RSI and GI are set at the early stages of project delivery and require significant effort to change. The policy indicators RAI, RRI and FSI allow for greater flexibility (under conditions) throughout the lifetime of the project.

Among the cases studied, only two projects, the Horgos-Nov Sad Motoway and the Attica Tollway, were limited by the implementation context, while for one, the A5 Maribor-Pince, the implementation context may be considered marginal. All other projects were delivered and operated under favourable implementation conditions. However, even under these positive conditions, not all projects realised their potential, and the respective TIRESI ratings predicted correctly the observed outcome. To this end, enhancement scenarios examined alternative contractual structures that included the integration of design and construction, a change of risk allocation, as well as the introduction of performance guarantees and penalties.

As observed in Table 5.21, all enhancement efforts targeted improvements in the GI and CSI, as issues related to less than anticipated project performance were identified to date back to the tendering and contractual set up of the respective projects. The A5 Maribor –Pince and the Attica Tollway attained positive outcomes, as both demonstrated supportive values of the GI and CSI.

It should be noted that apart from one scenario considered for the Reims Tramway project case, all other scenarios are hypothetical, as these could only have been implemented at the planning and procurement stage. This is to be expected, as they focus on the GI and CSI, which are set at the award stage and may change only under specific conditions (e.g. renegotiations, flexible contract).

The Reims Tramway case also offered a scenario that would enhance the project's exclusivity during operation (essentially aiming to improve the RSI), which provided an improved likelihood of reaching traffic and revenue targets. The proposed scenario is to be applied following recent renegotiations.

Finally, the E39 Klett–Barshaug, a well-structured PPP project delivered and operated under favourable implementation conditions has a financing scheme reflected in FSI = 0.719 (note: FSI = 1.000 depicts a public financed project). The enhancement scenario tested the potential of increasing the private contribution (FSI = 0.506). The TIRESI continued to predict a positive outlook with respect to the likelihood of reaching project targets. It should be noted that this alternative is available to decision makers. However, the decision is political and expresses the desirability of increasing private involvement in infrastructure delivery.

In conclusion, the presentation of these nine project cases and their proposed scenarios of improvement, confirmed the importance of the project structure and the fact that, in practical terms, only actions related to the policy indicators are possible following contract award and completion of construction.

Furthermore, the project cases also illustrated that the implementation context, especially, the FEI, while important, is not decisive of the project's potential of attaining project outcomes.

5.11.2 'What if' scenarios

The TIRESI rating system allows for alternative scenarios to be tested regarding project delivery. Projects may also be tested as per their performance under adverse financial economic conditions. Therefore, 'what if' scenarios were also tested.

Three project cases addressed such alternatives:

- The A5 Maribor–Pince Motorway project, delivered through public financing, showed a positive rating (likelihood) of achieving outcomes if it were delivered as a PPP. In this case, the Financing Scheme Indicator was reduced to FSI = 0.620 from FSI = 1.000, the latter reflecting public financing.
- The Attica Tollway demonstrated considerable resilience to the economic crisis. It is delivered with an FSI = 0.561 (i.e. with investors/sponsors anticipating considerable returns). The 'what if' scenario tested aimed at reducing the cost of capital by introducing crowdfunding. If similar conditions existed today as during the award stage (1997), crowdfunding could have presented an opportunity. Notably, crowdfunding was not a promoted financing option in 1997. With respect to the TIRESI rating, the cost and time to completion outcome rating of the project is retained at a lesser cost of capital.
- The E39 Klett–Barshaug provides evidence of a resilient internal structure even if, hypothetically, the macro-economic environment worsens (FEI drops to values below 0.60). A low FEI only slightly reduces rating scores for all project outcomes, while the time, traffic and revenue outcomes maintain their positive 'A' rating.

5.11.3 Limits and range of application

The TIRESI methodology and rating system was not able to provide ratings for airport, seaport and rail infrastructure. However, the key characteristics of each mode for which a rating system is available have been demonstrated. In this context, two cases were rated: the Agaete seaport and the Combiplan road and rail project. The Agaete seaport serves the connection between Gran Canaria and Tenerife resembling a fixed link between two previously isolated locations. The Combiplan does not include all the complexities of a rail project as it only concerns the development of a 6 kilometres section resembling a link (road). Hence, these two projects were addressed as a bridge/tunnel and a road, respectively. The ability of the TIRESI to predict observed performance with respect to outcomes justified the application.

The Brabo 1 tramway project demonstrated positive performance on all outcomes, with minimum divergence from expected outcomes. The TIRESI rating was more conservative. Notably, the GI was slightly lower than its threshold value for urban transit projects. More specifically, GI = 0.688 compared to GI = 0.700, which is the respective threshold for the urban transit infrastructure mode. A slight increase in the GI provided ratings that were in line with performance. Notably, as it was stated in Chapter 4, threshold values should be considered fuzzy rather than absolute and respective assessments taking into consideration this parameter should be made in cases for which indicator values are near to their thresholds.

5.11.4 Concluding remarks

The nine cases discussed in this chapter demonstrated the applicability of the TIRESI rating system. However, at the same time, through the various alternative scenarios and enhancement measures proposed for each case, it becomes evident that:

- The implementation context is important in infrastructure project delivery, but project performance with respect to the attainment of target outcomes depends significantly on the structure of the project (structural indicators). Though this observation, the importance of the procurement and contracting project phase is once again emphasised.
- Once the project is contracted, changing the structural elements of a project becomes less feasible.
- Policy indicators may change during project operation. However, as policy indicators are changed by inducing changes to the financing and funding scheme, these measures entail policy decisions concerning the project.

Stemming from the latter remark, it is implied that not only is project success related to the significance each stakeholder may attach to the achievement of any of the four outcomes discussed in this book, but in addition, it is related to the possible trade-offs that would need to be considered.

Stakeholder views are further discussed in the next Chapter 6.

5.12 Note

1   In spite of the license, *Naviera Armas* is not providing any services heretofore.

5.13 References

Attiki Odos Contract, Greek Law 2445/1996 of the Official Gazette.
Attiki Odos Official Website: www.aodos.gr.
Auto*Mat (2009). Dejme tunelu Blanka smysl. http://tunel-blanka.cz/chci-pomoci/co-chce-automat (Accessed 5 May 2015).

Beheersmaatschappij Antwerpen Mobiel. (2013). Public Transport: Redevelopment of City Boulevards and Extension of Tram Lines (BRABO 1). www.bamnv.be/Public_transport/2624/bam (Accessed 28 January 2013).

Bělohlav, Jiří (2015). 'Šéf Metrostavu: Tunel Blanka se může rozjet, ale jako by to někdo nechtěl.' Interviewed by B. Pečinka in: *Reflex*, published 11 March 2015. www.reflex.cz/clanek/rozhovory/62771/sef-metrostavu-tunel-blanka-se-muze-rozjet-ale-jako-by-to-nekdo-nechtel.html (Accessed 15 May 2015).

Bjørberg, S., Kristiansen, B.F., Graham, L., and Temeljotov, A.S. (2014). E-39 Klett-Bardshaug. In Roumboutsos, A., Farrell S., and Verhoest, K. (2014). *COST Action TU1001 – Public Private Partnerships in Transport: Trends & Theory: 2014 Discussion Series: Country Profiles & Case Studies*, ISBN 978-88-6922-009-8, pp. 133–141.

Bonnet, G., and Chomat, G. (2013). *The Reims' Tramway*. In Roumboutsos, A., Farrell, S., Liyanage, C. L. and Macário, R. (2013). *COST Action TU1001 Public Private Partnerships in Transport: Trends & Theory P3T3, 2013 Discussion Papers Part II Case Studies*, ISBN 978-88-97781-61-5, pp. 208–217.

Carpintero, S. (2009). Toll roads in Central and Eastern Europe: Promises and Performance. *Madrid: Transport Reviews*, 30(3), pp. 337–359.

CERTU. (2012). *Transports Public Urbains en France – Organisation Institutionnelle*. Lyon: CERTU.

Chen, L., and Manley, K. (2014). Validation of an Instrument to Measure Governance and Performance on Collaborative Infrastructure Projects. *Journal of Construction Engineering and Management*, 140(5).

Chen, L., Manley, K., and Lewis, J. (2012). Exploring governance issues on collaborative contracts in the construction industry. *Proceedings of the International Conference on Value Engineering and Management: Innovation in the Value Methodology*, The Hong Kong Polytechnic University, Hong Kong, pp. 65–70.

COWI Consulting Engineers and Planners AS. (2003). Pre-Feasibility Study of Second Carriageway Lane Construction on E-75 Highway Stretch: Hungarian Border (Horgos) – Novi Sad L=108,00 km. Belgrade, March 2003.

DARS. 2014. Obremenjenost cest (road congestion). www.dars.si/Dokumenti/O_avtocestah/Prometne_obremenitve/Obremenjenost_cest_97.aspx?print=1.

De Rus, G. (1997). How Competition Delivers Positive Results in Transport-A Case Study, The World Bank Group Viewpoint Note No. 136., Washington DC.

Frey, C. (2015). Moins de bus pour le même coût dans l'agglomération rémoise. *L'Union*. 4 February 2015. www.lunion.fr/region/moins-de-bus-pour-le-meme-cout-dans-l-agglomeration-remoise-ia3b24n484469 (4 Accessed February 2015].

Halkias, B., Roumboutsos, A., and Pantelias, A. (2013). Attica Tollway. In Roumboutsos, A., Farrell, S., Liyanage, C. L. and Macário, R. (2013). *COST Action TU1001 Public Private Partnerships in Transport: Trends & Theory P3T3, 2013 Discussion Papers Part II Case Studies*, pp. 28–38, ISBN 978-88-97781-61-5.

Halkias, B., and Tyrogianni, E. (2009). PPP Projects in Greece: The Case of Attica Tollway. Routes/Roads PIARC, April 2009.

Halkias, B., Tyrogianni, E., and Kitsos, D. (2008). A Significant Infrastructure Project within the Urban Environment of Athens: The Case of Attica Tollway. IABSE September 2008.

Harito, J., and Morello, S. (2011). Performance Plus. *ITS International*, 17(3), pp. 44–45.

Hojs, A., Liyanage, C., and Temeljotov, A.S. (2012). Analysis of Critical Success Factors for PPP Road Projects in Slovenia. In Michell, K., Bowen, P., Cattell, K. (eds.). *Delivering Value to the Community: Proceedings*. Cape Town: Department of Construction Economics and Management, pp. 448–454.

Kappeler. A., and Nemoz, M. (2010). Public-Private Partnerships in Europe-Before and during the Recent Financial Crisis. Economic and Financial Report 2010/04, European Investment Bank (EIB).

Koridori Srbije. Official Presentation. Project North. www.koridor10.rs/en/project-north (Accessed 15 December 2014)

Law (2003). Ley de puertos de Canarias. www.gobiernodecanarias.org/libroazul/pdf/39551. pdf.

Léko, I. (2014). 'Tunel Blanka, kauza Opencard a „elegantní" řešení pro všechny'. Česká pozice. 21 November 2014 (Accessed 15 May 2015).

Li, H., Arditi, D., and Wang, Z. (2012). Factors That Affect Transaction Costs in Construction Projects. *Journal of Construction Engineering and Management*, 139(1), 60–68.

Łukasiewicz, A., Roumboutsos, A., Bernardino, J., Brambilla, M., Liyanage, C., Mitusch, K., Mladenovic, G., and Pantelias, A. (2015). Deliverable D2.1 – BENEFIT Database, BENEFIT (Business Models for enhancing Funding and enabling Financing for Infrastructure in Transport) Horizon 2020 Project, Grant Agreement No 635973. www. benefit4transport.eu/index.php/reports.

Mandalozis, D., Halkias, B., Tyrogianni, H., and Kalfa, N. (2012). The Carbon Footprint of Attica Tollway. TRA-Europe 2012. *Procedia – Social and Behavioral Sciences*, 48, pp. 2988–2998.

Mestskyokruh.info (2007). 'Přínosy a důsledky výstavby městského okruhu.' mestskyokruh. info/mestsky-okruh/vychodni-cast-mo-libenska-spojka/prinosy-dusledky-vychodni-cast-mo-libenska-spojka/celospolecenske-prinosy-vychodni-cast-mo-libenska-spojka/ (Accessed 15 May 2015).

Mladenović, G., Roumboutsos, A., Campos, J., Cardenas, I., Cirilovic, J., Costa, J., González, M.M., Gouin, T., Hussain, O., Kapros, S., Karousos, I., Kleizen, B., Konstantinopoulos, E., Lukasiewicz, A., Macário, R., Manrique, C., Mikic, M., Moraiti, P., Moschouli, E., Nikolic, A., Nouaille, P.F., Pedro, M., Inchausti-Sintes, F., Soecipto, M., Trujillo Castellano, L., Vajdic, N., Vanelslander, T., Verhoest, C., and Voordijk, H. (2016). Deliverable D4.4- Effects of the Crisis & Recommendations, BENEFIT (Business Models for enhancing Funding and enabling Financing for Infrastructure in Transport) Horizon 2020 project, Grant Agreement No. 635973. www.benefit4transport. eu/index.php/reports.

Newspaper 'Blic'. Route of Corridor 10 Will Be Built by Serbian Consortium [online]. www.blic.rs/Vesti/Ekonomija/184687/Trasu-Koridora-10-gradice-konzorcijum-srpskih-preduzeca (Accessed 17 December 2014).

Odeck, J. (2014). E-18 Grimstad-Kristiansand. In Roumboutsos, A., Farrell S., and Verhoest, K. (2014). *COST Action TU1001 – Public Private Partnerships in Transport: Trends & Theory: 2014 Discussion Series: Country Profiles & Case Studies*, pp. 142–148, ISBN 978–88–6922–009–8.

Organic Law. (1996). De reforma de la Ley orgánica 10/1982 del Estatuto de Canarias. https://www.boe.es/buscar/doc.php?id=BOE-A-1996-29114.

Oživení. (2012). Otevřenost zadávacích řízení v ČR. www.bezkorupce.cz/faqs/co-je-to-zadavaci-rizeni-jake-jsou-druhy-zadavaciho-rizeni-a-jaky-je-mezi-nimi-rozdil/ (Accessed on 5 May 2015).

Pantelias, A., Roumboutsos, A., Bernardino, J., Bonvino, G., Cardenas, I., Dimitriou, H., Karadimitriou, N., Karousos, I., Kolahi, A., Mitusch, K., Moraiti, P., Moschouli, E., Trujillo, L., Vajdic, N., Vanelslander, T., Verhoest, K., Voordijk, H., Willems, T., and Wright, P. (2015). Deliverable D3.1 – Methodological Framework for Ex-post Analysis, BENEFIT (Business Models for enhancing Funding and Enabling Financing for

Infrastructure in Transport) Horizon 2020 Project, Grant Agreement No. 635973. www.benefit4transport.eu/index.php/reports.

Papaioannou, P. (2006). Recent Experience on Success and Failure Stories from Funding Large Transportation Projects in Greece, *1st International Conference on Funding Transportation Infrastructure, Banff*, Alberta, Canada, 2–3 August 2006.

Papandreou, K., and Tyrogianni, E. (2007). Level of Service in Concession Motorway Projects. *XXXV ASECAP Study and Information Days*, Crete, 2007.

Prague City Hall. (2015). Rozpočet hl.m. Prahy 2015 (Prague City budget 2015). www.praha.eu/jnp/cz/o_meste/magistrat/tiskovy_servis/tiskove_zpravy/rada_hl_m_prahy_odsouhlasila_navrh.html (Accessed 25 May 2015).

Prague City Hall. (2016). Blankou za rok projelo 30 milionu aut (Based on Interim Evaluation of Blanka Tunnel Complex's Impact on Prague Transport System). www.praha.eu/jnp/cz/o_meste/magistrat/tiskovy_servis/tiskove_zpravy/blankou_za_rok_projelo_30_milionu_aut.html.

PricewaterhouseCoopers. (2005). *Delivering the PPP Promise: A review of PPP Issues and Activity*. PricewaterhouseCoopers LLP. https://www.pwc.com/gx/en/government-infra structure/pdf/promisereport.pdf

Public Enterprise 'Roads of Serbia' homepage (online). Available at: www.putevi-srbije.rs/index.php?lang=sr&Itemid=2754 [Accessed 15 December 2014].

Reims Métropole. (2007). *Dossier d'enquête préalable à la déclaration d'utilité publique*.

Rijkswaterstaat. (2012). Standaardcontract DBFM. Den Haag, RWS.

Roumboutsos, A. (2015). *Business Models for Enhancing Funding and Enabling Financing for Infrastructure in Transport: PPP and Public Transport Infrastructure Financing Case Studies*. Horizon 2020 European Commission. Department of Shipping, Trade and Transport, University of the Aegean, Greece, ISBN 978-618-82078-1-3. www.benefit4transport.eu

Roumboutsos, A., Bange, C., Bernardino, J., Campos, J., Cardenas, I., Carvalho, D., Cirilovic, J., González, M.M., Gouin, T., Hussain, O., Kapros, S., Karousos, I., Kleizen, B., Konstantinopoulos, E., Leviäkangas, P., Lukasiewicz, A., Macário, R., Manrique, C., Mikic, M., Mitusch, K., Mladenovic, G., Moraiti, P., Moschouli, E., Nouaille, P.F., Oliveira, M., Pantelias, A., Inchausti-Sintes, F., Soecipto, M., Trujillo, L., Vajdic, N., Vanelslander, T., Verhoest, K., Vieira, J., and Voordijk, H. (2016). Deliverable D4.2-Lessons Learned–2nd Stage Analysis, BENEFIT (Business Models for Enhancing Funding and Enabling Financing for Infrastructure in Transport) Horizon 2020 Project, Grant Agreement No. 635973. www.benefit4transport.eu/index.php/reports.

Roumboutsos, A., Farrell, S., Liyanage, C. L., and Macário, R. (2013) COST Action TU1001 Public Private Partnerships in Transport: Trends & Theory P3T3, 2013 Discussion Papers Part II Case Studies, ISBN 978–88–97781–61–5.

Roumboutsos, A., Farrell S., and Verhoest, K. (2014). COST Action TU1001 – Public Private Partnerships in Transport: Trends & Theory: 2014 Discussion Series: Country Profiles & Case Studies, ISBN 978–88–6922–009–8.

Roumboutsos, A., Pantelias, A., Sfakianakis, E., Edkins, A., Karousos, I., Konstantinopoulos, E., Leviäkangas, P., and Moraiti, P. (2016). Deliverable D3.2- The Decision Matching Framework Policy Guiding Tool, Project Rating Methodology and Methodological Framework to Increase Business Model Creditworthiness, BENEFIT (Business Models for Enhancing Funding and Enabling Financing for Infrastructure in Transport) Horizon 2020 Project, Grant Agreement No. 635973. www.benefit4transport.eu/index.php/reports.

Royal Decree. (2250/1985). Sobre traspaso de funciones y servicios de la Administración del Estado a la Comunidad Autónoma de Canarias en Materia de Puertos. www.boe.es/diario_boe/txt.php?id=BOE-A-1985-25096.

Sofianos, A.I., Loukas, P., and Chantzakos, C. H. (2004). Pipe jacking a sewer under Athens. *Tunnelling and Underground Space Technology*, 19(2), pp. 193–203.

Susarla, A. (2012). Contractual Flexibility, Rent Seeking, and Renegotiation Design: An Empirical Analysis of Information Technology Outsourcing Contracts. *Management Science*, 58(7), pp. 1388–1407.

Tyrogianni, H., Halkias, B. Politou, A., and Kotzampassi, P. (2012). The Attica Tollway Operations Authority KPI Performance System. TRA-Europe 2012. *Procedia – Social and Behavioral Sciences*, 48, pp. 2999–3008.

Vajdic, N., and Mladenovic, G. (2013). Horgos-Pozega, Toll Motorway Concession, Serbia, in *COST Action TU1001 – 2013 Discussion Papers: Part II Case Studies*, Edited by Athena Roumboutsos, Sheila Farrell, Champika Lasanthi Liyanage and Rosário Macário, COST Office, pp. 47–54, ISBN 978–88–97781–61–5.

Vanelslander, T., Roumboutsos. A., Voordijk, H., Cardenas, I., Bernardino, J., Carvalho, D., Vanelslander, T., Moschouli, E., Verhoest, K., Williems, T., Moraiti, P., and Karousos, I. (2015). Deliverable D2.2- Funding Schemes & Business Models, BENEFIT (Business Models for Enhancing Funding and Enabling Financing for Infrastructure in Transport) Horizon 2020 Project, Grant Agreement No. 635973. www.benefit4transport. eu/index.php/reports.

Voordijk, H., Roumboutsos, A., Cardenas, I., Lukasiewicz, A., Macário, R., and Pantelias, A. (2015). Deliverable D2.4 – Governance Typology, BENEFIT (Business Models for Enhancing Funding and Enabling Financing for Infrastructure in Transport) Horizon 2020 Project, Grant Agreement No. 635973. www.benefit4transport.eu/index.php/ reports.

Voordijk, J.T., Roumboutsos A., Bernardino, J., Brambilla, M., Cardenas, I., Gouin, T., Gremm, C., Inchausti Sintes, F., Karousos, I., Liyanage, C., Łukasiewicz, A. Macario, R., Mikic, M., Mitusch, K., Mladenovic, G., Moschouli, E., Nouaille, P., Torta, F., Trujillo, L., Vanleslander, T., Verhoest, K., Vieira, J., and Villalba-Romero, F. (2016). Deliverable D5.1 – Potential of Investments in Transport Infrastructure, BENEFIT (Business Models for Enhancing Funding and Enabling Financing for Infrastructure in Transport) Horizon 2020 Project, Grant Agreement No. 635973. www.benefit4transport. eu/index.php/reports.

White&Case (2011). 'Zpráva o právní provĕrce smluvní dokumentace k výstavbĕ tunelu Blanka' ('Report from the legal audit of contracts concerning tunnel Blanka'). www.praha.eu/public/48/37/a1/1142985_153973_Tunel_Blanka___Red_Flag_Report. pdf (Accessed 5 May 2015).

6 Investing in transport infrastructure

A stakeholder's view

Aristeidis Pantelias and
Athena Roumboutsos

6.1 Introduction

Investment in infrastructure does not take a single form. It can be divided into two major categories, namely investing in 'infrastructure as a business' and investing in 'infrastructure as an asset class'. Both investment perspectives carry their own goals, are based on their own incentives, involve their own uncertainties, and very significantly, pertain to different actors or stakeholders.

Investing in 'infrastructure as a business' is the most 'traditional' form of engagement with the overall sector. It pertains to the business of developing, constructing and operating infrastructure assets and the stakeholders involved are the ones that are seen to be active in the various industries related to the built environment, that is public sector agencies,[1] construction firms, architectural and planning firms, asset operators, financiers, etc. Investment in 'infrastructure as a business' is fundamentally and inherently sector-specific and in certain cases asset-specific, as stakeholders need to understand and have expertise on how projects need to be designed, built and operated, including also the available options for procuring and financing them.

Investment in 'infrastructure as an asset class' takes a much different perspective. It concentrates on the characteristics of the investment as a source of financial returns irrespective of the underlying infrastructure asset. In this case, actors, that is financial investors, are solely interested in the financial characteristics of the cash-flows that can be produced and captured by these assets, that is their underlying funding schemes, out of which their respective returns will be carved.

Traditionally, the actors involved in 'infrastructure as a business' were also the ones reaping the benefits of investing in 'infrastructure as an asset class'. In a sense, the involvement of private finance in infrastructure started with construction companies evolving into infrastructure developers, in charge of the entire life cycle of the asset and responsible for securing the necessary financing for its construction, operation and maintenance. This transition involved a fundamental change in industry mentality and structure that happened through various cycles of trial and error, as project stakeholders, despite their previous experience in procuring and delivering projects, were not properly equipped to manage this new type of 'business' and understand the characteristics of their investment. As the investor base interested in the financial returns generated by infrastructure has broadened over the last couple of decades, additional stakeholders have come on board such projects, with less expertise in the sector than those who were traditionally involved in it.

At the same time, investing in infrastructure is not an exact science and involves making a series of decisions under conditions of uncertainty. Often these decisions are made based on assessments of trade-offs whose ramifications are not always fully captured or understood beforehand for various reasons.

Inevitably, the high degree of complexity of infrastructure projects and the multiple externalities, positive and negative, generated by these investments necessitate the involvement of all or most stakeholders in the relevant decisions. As their interests and goals may be affected by project characteristics in different ways, both positively and adversely, further complexity is added to the decision making process.

Notably, the process is complicated even further by the fact that relevant information is not readily available to all stakeholders and even if it were, not all of them have the required expertise and knowledge background to properly process it. Finally, although project outcomes and their potential level or degree of achievement are usually clearly understood concepts, stakeholders may value them in different ways. They would thus attach a different weight to the attainment of a specific outcome depending on their individual interests, objectives and perspectives, adding one more layer of complexity to the decision-making process.

To date, various decision-support systems have been introduced in support of infrastructure investment decisions. These systems address and pertain to various stages of infrastructure development and their multiplicity is evidence of both the complexity of the assessment as well as the perceptions attached to the desirability of different possible outcomes. For example, the early recognition of the importance of front-end planning in the construction industry introduced the Project Definition Rating Index (PDRI) as early as 1994, while numerous rating systems have been developed to address infrastructure performance with respect to sustainability. In addition, addressing issues of investment reliability with a particular emphasis on debt repayment, project creditworthiness ratings are increasingly being propagated among the decision tools available to stakeholders.

This chapter aims to position the TIRESI and its rating system among the universe of systems and tools available to project stakeholders. While the scope of the TIRESI extends only to transport infrastructure projects, the aim is to identify and present key complementarities and overlaps. Notably, while it is widely known that the implementation phase of a project is central in achieving its outcomes, there has been no previous effort or current tool available that aims to capture the profile of this phase of project development and its impact on the likelihood of reaching pre-specified project outcomes. In this sense, the TIRESI fills an existing assessment gap right from the outset.

Without doubt, there is merit in assessing the potential to achieve outcome targets through the proposed TIRESI ratings. As discussed previously, transport infrastructure projects target various project outcomes, the attainment of which can be considered as a measure of their success. However, success does not have the same meaning for all project stakeholders. Depending on their considerations, different stakeholders will place dissimilar value to the various possible outcomes that a project can achieve. This is only natural as projects do not have to secure a full alignment of interests from all participating parties, but just enough common ground that would be able to create the right incentives to move the project forward and safeguard its implementation considering the effect of potential changes (both internal as well as external) that may be incurred during its life-time.

In this context, this chapter considers the use of the TIRESI and its rating system from the point of view of different stakeholders involved in transport infrastructure projects. As an example, TIRESI ratings can assist project stake-holders in determining the nature and extent of their involvement in different project phases. For private parties, this assessment may be related to corresponding investment decisions, whereas for the public party, it can help to better allocate risks or assess the influence project structure decisions may have on the project's potential of reaching specific outcome targets under various implementation context conditions.

Finally, this chapter provides a more in-depth discussion on the O-TIRESI. This version of the TIRESI may be useful to less knowledgeable stakeholders as it aims to summarise information coming from the S-TIRESI and/or the D-TIRESI. The O-TIRESI may guide project-related decisions and its computation methodology is sufficiently flexible to accommodate different stakeholder value systems.

6.2 The Transport Infrastructure Resilience Indicator Rating System: a stakeholder's view

The TIRESI rating system assesses performance following project award. At this point many project decisions have already been made and conditions set. These include, for example, infrastructure type, size of investment, location as well as the delivery model (fully public or including private financing). Based on studies conducted before the tendering stage, that is pre-feasibility, feasibility, etc., the key target outcomes of a project have also been defined and set. With respect to this book, these targets include the following:

- construction budget;
- construction duration;
- anticipated level of traffic;
- anticipated level of revenues.

Additionally, the discussions in Chapters 2 and 3 have highlighted the following:

- the high influence of the implementation context on project performance, with the InI representing the level of the project's implementation stability;
- the impact of decisions made at the procurement phase on a project's ability to reach its targets;
- the limited ability to enhance performance after the procurement phase, which is then mostly focused on innovation and business development opportunities;
- the role that funding and Financing Scheme Indicators may play in driving project outcomes under specific conditions already set.

Within this context, by using the TIRESI and its underlying rating system in an ex-ante approach, project structuring scenarios may be considered. When assessing these scenarios, the target would be to accomplish a B_{EX}^+ TIRESI rating for all figures of merit of interest, if possible. As explained previously, a B_{EX}^+ rating would secure a rather high likelihood of attaining a specific figure of merit target even under relatively adverse implementation conditions.

Effectively, assessing different project structuring scenarios would be relevant to all phases of project delivery, namely (in chronological order):

- The planning phase: where alternative decisions of project design and structure may be tested under different procurement options (PPP model or public financing), implementation conditions as well as funding and financing schemes. At this point, new financing instruments can also be tested as well as different variations for the potential capital structure of the project. Notably, at this stage decision makers can identify the minimum procurement expectations.
- The procurement phase: by defining the minimum number of bidders per tender; the minimum requirements for bidder selection; the minimum contractual conditions and other issues related to the governance (GI), cost saving (CSI) and revenue support (RSI) indicators.
- Financial Close phase: where the expected range of values of different variations of the FSI supporting the selected scenario may guide the specification of the capital structure of the project. The contribution of new and innovative financing instruments may be assessed with respect to the selected scenario of project implementation.
- The implementation phase: by investigating potential opportunities of including innovation and new business activities, and improving the capabilities of the contracting authority in terms of monitoring and management.
- Renegotiations phase (if applicable): given the project's structure and implementation context, decisions during potential future renegotiations may be supported and guided toward attaining indicator values that will support the achievement of specific project outcome targets, especially with respect to modifying the project's funding and financing schemes.

Finally, employed as a monitoring instrument, the TIRESI may assist in providing future predictions of the likelihood of attaining certain project performance targets and in adopting measures during the entire duration of a project's life cycle that minimise the adverse effects of unforeseen changes (both internal and external) or capitalise on other relevant opportunities.

It is interesting to note that all the above potential assessments that could take place at different stages of a project's life cycle may be relevant to both types of infrastructure investors, that is those interested in the 'business' of delivering infrastructure and those interested in the relevant 'financial returns'. Arguably, business investors, being naturally closer to the asset and its needs, would look for a deeper level of understanding and a wider range of sensitivities and analyses to ascertain that their business objectives will be met. Financial investors, however, would also need to develop a level of comfort with many project parameters and sensitivities, especially those affecting revenue streams and thus their expected returns. In a nutshell, the analyses that the TIRESI could support per project phase could add value to the decision-support process of all types of investors involved by helping them clarify and understand various aspects of the project's development that are relevant to their individual interests and objectives.

Notably, the caveat is that all the aforementioned investigations and analyses are not always feasible or politically desirable as demonstrated in the project case studies of Chapter 5.

It should be stressed, that TIRESI ratings are not aiming in any way to substitute or assess the validity of the original method(s) or underlying assumptions used for the definition of the various outcome target values. It is explicitly assumed that project outcome targets have been carefully and properly estimated and researched during the planning stage of the project using appropriate estimation tools and expertise. The TIRESI ratings aim to assess whether the structure of the project, as being delivered and within its implementation context, bolsters or hinders the attainment of the specified targets.

In this context, TIRESI ratings provide decision makers with information on the potential to achieve planned objectives based on how a project is implemented and can be meaningless or misleading in cases where unrealistic project targets have been set. For example, if a project's budget has been underestimated (or overestimated), the TIRESI rating will not be able to reflect this and may provide an A rating for the respective figure of merit (cost to completion).

Conversely, an A rating is assigned to projects demonstrating both exogenous and endogenous resilience. Such a rating reflects, among other things, strong and mature institutions underlying the project's implementation context as well as a high level of competence of the actors involved in delivering the project. Implicitly then, this rating suggests, despite being qualitative in nature, that several necessary pre-conditions already exist in order to set and pursue realistic project objectives at the planning stage. In a similar vein, a rating C corresponds to poor endogenous (but not necessarily and always to poor exogenous) project resilience, that is a poorly structured project during implementation. In this case, even if the project has well-justified and researched project outcome targets, its internal project structure is not supportive of their attainment.

Based on the above observations, projects assigned to the A and C rating categories are expected to have greater accuracy in the reflected likelihood to attain their outcome targets. The B rating category, however, as it takes the middle ground and covers a lot of 'grey' areas, is more reliant on the accurate estimation of project outcomes in the planning phase of the project.

6.2.1 Overall Transport Infrastructure Resilience Indicator (O-TIRESI)

Originally defined in Chapter 4, the O-TIRESI is typically provided as a set comprising four ratings, each one corresponding to one of the four figures of merit (outcomes) considered by the methodology. The sequence of the four ratings takes the form of:

$$0 - TIRESI = \{Rating_{Cost}, Rating_{Time}, Rating_{Traffic}, Rating_{Revenues}\}$$

where:

$Rating_{Cost}$ is the rating describing likelihood of reaching cost to completion target.

$Rating_{Time}$ is the rating describing likelihood of reaching time to completion target.

$Rating_{Traffic}$ is the rating describing likelihood of achieving forecast traffic.

$Rating_{Revenues}$ is the rating describing likelihood of achieving forecast revenue.

Such a presentation of an overall rating can be of essence to some stakeholder groups and in this context, this section aims to provide guidelines as to how the O-TIRESI may be adjusted to their specific needs.

In order to quantify the O-TIRESI, three different approaches are possible:

- The O-TIRESI is equal to its 'weakest link', that is it is equal to the minimum of the ratings describing the four figures of merit (outcomes) considered in the project of interest. However, in this approach, the potential importance or ability of the 'strongest link' to compensate for shortcomings is overlooked.
- The O-TIRESI is equal to a weighted average of the four figures of merit considered in the project of interest. This approach needs to be tailored to the needs of each stakeholder as the respective weighting applied would have to reflect the individual stakeholder's risk perceptions and interests. Notably, in this approach, stakeholders might explicitly focus on one specific outcome of interest by assigning a weighing of zero (0) to all other outcomes.
- The O-TIRESI is simply the set comprising the four figures of merit. This is the 'default' approach as all it does is present the four ratings side by side without any further processing.

All approaches bear advantages and disadvantages and may be equally applied using either the S-TIRESI or the D-TIRESI ratings. Combinations of the approaches may also be considered as any of them may be applied per project phase.

Notably, if all figures of merit (outcomes) have obtained an equal rating then the process is further simplified. If the S-TIRESI (or D-TIRESI) rating, however, varies between outcomes, then the O-TIRESI would need to express the decision maker's interests and viewpoints.

As mentioned earlier, the overall TIRESI methodology aims to remain stake-holder-neutral, that is it is not assuming the point of view of any stakeholder. Therefore, the estimation of the O-TIRESI is ultimately left to each stakeholder to consider based on the alternative approaches presented. Consequently, different stakeholders may end up with different overall assessments for the same project based on their own value system. In this context, none of the projects in the BENEFIT project case dataset were judged as successful or not. They were simply assessed with respect to achieving, over-achieving or under-achieving their individual outcome targets.

Table 6.1 presents examples of potential applications. It should be stated that the TIRESI rating system is qualitative in nature and at its current level of development, the precise likelihood of each rating category cannot be assessed as a per cent (%) probability. The TIRESI rating scale ranges from C_- to A^+ and can indicatively be mapped to a numerical scale of likelihoods taking values within the range [1,9]. This range is adopted in the estimations presented in Table 6.1, which illustrate all possible alternative quantification approaches of the O-TIRESI from the ratings of the four figures of merit (cost and time to completion, actual versus forecast traffic and revenues). The example concerns two projects presented in Chapter 5, namely the E39 Klett–Bardshaug road project and the Blanka tunnel project. For both projects, the ratings of all figures of merit correspond to the S-TIRESI and have been determined based on indicator values from their snapshots at award.

The information in Table 6.1 can be used by different stakeholders in different ways. For example, a construction contractor, interested in delivering the asset within cost and budget would probably avoid merging the attainment of these targets with operational characteristics such as traffic and revenue. In their case, the 'Default' approach would be able to inform them about the vulnerability of their objectives by considering simply the first two ratings within the provided sets. A budget-constrained public authority that wants to maintain a high degree of accountability towards its public constituents may wish to use the weighted average approach by adding more emphasis in delivering the project under budget and ensuring that all other targets are met to their respective degrees. Finally, a financial investor (e.g. a pension fund manager) aiming to allocate part of their investment portfolio on infrastructure assets may want to consider the 'weakest link' approach to make sure that the project they will be investing in is as safe as possible. In their case, they may have in mind an internal decision-making threshold that could qualify projects above an A rating and disqualify any project below it, based on their risk-return appetite.

Table 6.1 Examples of the Overall Transport Infrastructure Resilience Indicator assessment (O-TIRESI)

Project		Cost-to-compl.	Time-to-compl.	Actual vs forecast traffic	Actual vs forecast revenue	O-TIRESI
Approach: weakest link						
E39 Klett-Bardshaug	S-TIRESI	A^+	A	A	A^+	A
Blanka Tunnel	S-TIRESI	B_{EN-}	C^+	C^+	C_-	C_-
Approach: Weighted Average						
Example of Stakeholder's Weighting						
E39		40%	10%	20%	30%	
Klett-Bardshaug	S-TIRESI	A^+	A	A	A^+	A^+
	Numerical Mapping	9	8	8	9	
	Weighting	0,4*9	0,1*8	0,2*8	0,3*9	
Blanka Tunnel	S-TIRESI	B_{EN-}	C^+	C^+	C_-	C^-
	Numerical Mapping	4	3	3	1	
	Weighting	0,4*4	0,1*3	0,2*3	0,3*1	
Approach: default						
E39 Klett-Bardshaug	S-TIRESI	A^+	A	A	A^+	A^+,A,A,A^+
Blanka Tunnel	S-TIRESI	B_{EN-}	C^+	C^+	C_-	B_{EN-},C^+,C^+,C_-

Source: Authors' compilation.

6.3 Comparing the TIRESI with infrastructure decision-support tools: common grounds

As briefly discussed in the introduction of Chapter 6, the need to support and guide decision makers has led to the development of various decision-support systems that address particular aspects of project development and delivery. These systems stem from the systematic consideration of best practices and lessons learnt, which is then formulated into check lists or standardised assessments many times in the form of ratings of the likelihood that a given activity, process or element in project delivery has been successfully addressed. These assessments and ratings are meant to provide market professionals with quality information to form a solid base for decision making.

The present section, seeks to position the TIRESI rating among other similar and popular decision support systems. The discussion does not aim to be exhaustive but rather indicative, as summarising all available tools is outside the scope of this book. The focus, therefore, has been on systems that are or can be related to transport infrastructure delivery, as this is the scope of the TIRESI. In this respect, three such systems (including groups of systems) are presented and discussed in relation to the TIRESI rating.

6.3.1 *Project Definition Rating Index for Infrastructure (PDRI-Infrastructure)*

Poor project outcomes may frequently be traced back to the planning stage and, therefore, the need to improve front-end planning processes has been emphasised (Atkinson *et al.*, 2006). An example of a front-end planning (FEP) rating index is provided by the Construction Industry Institute (CII). Its members extensively studied, documented and utilised the concept and produced rating indices supporting the FEP of industrial projects and buildings (CII, 1994, 2008a, 2008b) and thereafter of infrastructure projects (CII, 2010).

The rating is based on a scoring sheet in which a weighting is assigned based on the relative contribution to the risk of the specific element scored. The weighting system has been produced based on expert knowledge with respect to the experts' assessment of criticality assigned per item/process listed in the score sheet. The rating activity is mostly addressing project participants on their assessment of the status of project development process. The process of rating is expected to create a consensus view of the current level of planning of each of the elements, as well as of the overall project. Scoring is undertaken individually by project participants and these ratings are incorporated into a weighted model that provides a measure of the quality of the planning. The aggregation indicates the participants' assessment of project scope definition and highlights planning gaps, but also the differing perceptions, allowing participants to streamline their views (Bingham and Gibson, 2017).

The broad categories of elements that are considered in the PDRI-Infrastructure are compared to the elements considered by the TIRESI. Table 6.2 provides this

Table 6.2 Prominent elements of the PDRI-Infrastructure compared to TIRESI considerations

#	*Most prominent elements of the PDRI-Infrastructure*	*Considered in TIRESI through indicators (YES/NO)*
1	Need and purpose documentation	NO
2	Investment strategies and alternative assessments	YES (in RRI, RAI and FSI)
3	Contingencies	NO
4	Design and construction cost estimates	YES (as an outcome benchmark)
5	Design philosophy	YES (as level of technical difficulty & innovation in CSI)
6	Preliminary project schedule	YES (as an outcome benchmark)
7	Evaluation of compliance requirements	YES (in GI)
8	Existing environmental conditions	YES (as permits)
9	Capacity	NO (taken for granted)
10	Public involvement	YES (as contracting authority and its capability to manage stakeholders in GI)
11	Funding and programming	YES (in RRI and RAI)
12	Geotechnical characteristics	YES (as level of technical difficulty and knowledge of ground conditions in CSI)

Source: Authors' compilation.

comparison for the 12 elements considered most likely to affect the success of a project (Bingham and Gibson, 2017).

Notably, elements that were identified as bearing considerable impact on the potential to achieve outcomes and in many instances identified as the origin of poor performance (see Chapter 5, case examples), while included in the PDRI-Infrastructure scoring sheet, were not in the list of the 12 most prominent elements. Such an omission includes the procurement strategy, which to a large extent defines the GI and the expected competence(s) of the actors involved (as part of the CSI).

6.3.2 *Sustainability rating systems for infrastructure*

The concept of sustainable development is usually taken to encompass the need for finding an acceptable balance between the priorities of economic development, social progress and environmental protection. There is a significant history of international agreement on the need for sustainable development. The 1972 UN Conference on the Human Environment produced a declaration publicly acknowledging the adverse impact of humans on the natural world. Ever since, the notion of sustainability has gradually secured its position in the construction industry, as construction in general, and infrastructure, in particular, have significant environmental impacts. Sustainable development is usually studied by using three primary categories (triple bottom line): (i) social, (ii) environmental and (iii) economical and, therefore, concerns technical, environmental, economic, social and individual sustainability.

In support of providing managers and decision makers with assessment tools, numerous rating systems for infrastructure have been proposed and applied by the industry, with transport infrastructure being a late comer (Krekeler *et al.*, 2010). These may be classified into two groups: (i) qualitative tools based on scores and criteria and (ii) quantitative tools using a physical life-cycle approach with quantitative input and output data on flows of matter and energy (Forsberg and Malmborg, 2004). Furthermore, while sustainability is a wider concept, sustainability rating systems for infrastructure tend to focus on five categories: use of resources, energy, transport, water and waste. Clevenger *et al.* (2013) reviewed six sustainability rating systems commonly used for infrastructure projects (mostly applicable in the USA): BEST-in-Highways, Envision, GreenLITES, Greenroads, I-LAST and Invest. While the detailed description of each rating system is beyond the scope of the present discussion, it is worth noting that these rating systems are typically built on a combination of scorecards and self-assessment and third party review including weighted assessments and project award systems. Furthermore, there is considerable overlap between these systems and their key differentiation lies in the weights assigned to factors, as each system has different objectives per stakeholder and project.

The TIRESI rating system does not include a sustainability factor per se, at least as these are considered in the commonly used sustainability rating systems for infrastructure. However, for the key outcomes that are assessed: cost to completion; time to completion; actual versus forecast traffic and actual versus forecast revenues, the relevant targets, should be defined based on sustainability criteria.

6.3.3 Creditworthiness project ratings

Ratings of project creditworthiness concern the assessment of the likelihood of default on debt obligations or the delayed payment of debt. Various credit assessment methodologies currently exist for infrastructure projects that are rigorous, comprehensive and in use for quite some time, having been continuously scrutinised by their developers to gain the trust of capital market financiers. In the context of privately (co)financed infrastructure projects, these methodologies are directly connected to project risks and their evolution over the life cycle of the contractual agreement. To derive project credit assessments, full and in-depth knowledge of how risks are shared is required, along with all potential guarantees that may exist by and for all parties involved. The detailed presentation of these methodologies is beyond the scope of this book. For a review and critique, the interested reader can refer to Cantor and Parker (1994), Pantelias and Roumboutsos (2015) and Roumboutsos *et al.* (2016), among other relevant studies.

Credit assessment methodologies serve a specific purpose that is a lot more limited than what the TIRESI aims to accomplish. As mentioned previously, these methodologies are concerned with the ability of a project to service its debt obligations and therefore their focus is on risk bearing, that is on the (legal) entity that will be liable to honour (pay) these obligations when they fall due. The TIRESI places its focus on risk management that comes a step earlier than the ability to bear risk. The rationale is that various circumstances that can jeopardise the ability of a project to deliver its expected outcomes (and to repay its debt obligations as a sub-set of them) can be anticipated and potentially mitigated through managerial actions/decisions before the need exists for one of the project stakeholders to bear the financial consequences of a risk that has materialised. Finally, credit assessment methodologies are concerned with infrastructure projects that are privately (co)financed, that is projects in which part of the financing originates from the private sector, such as projects procured as PPPs. The TIRESI aims to be more comprehensive considering both privately (co)financed as well as publicly financed infrastructure projects. In that sense and considering these fundamental differences in their scope, the TIRESI aims to act in a complementary way to these existing methodologies rather than as a replacement.

6.3.4 Concluding remarks

From the brief review of these three decision-support rating tools (or groups of tools), it is clear that the TIRESI rating methodology acts in a complementary manner to them. Some of these tools focus exclusively on the front-end planning of infrastructure projects, thus ignoring the importance of the implementation phase (i.e. design, construction and operation) in achieving success. Sustainability rating systems focus on specific attributes of project development that are mostly concerned with the 'triple bottom line'. While their assessment extends from the planning phase to part of the implementation, their focus is too specific to be of interest to all project stakeholders at the same level of importance. Finally, credit assessment methodologies take a view on the entire implementation life cycle of a project, from construction to operation, but limit their focus on the ability of the project to honour its debt obligations. The TIRESI, while taking the outputs of the planning phase as given, focuses on many different aspects of project implementation aiming to capture the interrelations between various internal and external project elements that may impact its ability to deliver different outcome targets. In that sense, this review has provided additional evidence that the TIRESI fills an existing assessment gap that can provide project stakeholders with additional information and support in their relevant investment decisions. Figure 6.1 illustrates the complementarity of the various rating systems.

6.4 Conclusions

In this chapter, the TIRESI methodology was looked at under the lenses of different possible project stakeholders. As recognised from the outset, infrastructure

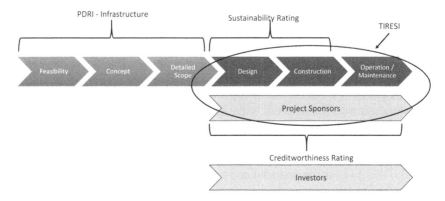

Figure 6.1 Complementarity of infrastructure rating systems

investors belong to two different major groups and may have different interests, objectives and perspectives with respect to expected project outcomes. The TIRESI can cater to these different needs as it has been developed to be stakeholder neutral. In that sense, the O-TIRESI, the version of the indicator, which aims to summarise information for all four project outcome targets considered, can be quantified by three different approaches, based on stakeholders' requirements. The TIRESI aims to fill an identified gap in project assessment. Various tools currently exist in the market, performing various functions and assessments and supporting different project phases and interested stakeholders. The TIRESI has been found to have minimum overlap with these existing tools, while contributing new information that is not provided by any of them. Although the TIRESI is applicable only in the sector of transport infrastructure, its assessments are rigorous and systematic and can ultimately add value by supporting the investment decisions of various stakeholders.

6.5 Note

1   In the case of public sector agencies, infrastructure 'as a business' is not meant in the sense of profit-generating private sector enterprise, but as the formal process of planning, designing, procuring and delivering infrastructure assets to increase social welfare and support economic development.

6.6 References

Atkinson, R., Crawford, L., and Ward, S. (2006). Fundamental uncertainties in projects and the scope of project management. *International Journal of Project Management*, 24(8), pp. 687–698.

Bingham, E., and Gibson, G.E., Jr. (2017). Infrastructure Project Scope Definition Using Project Definition Rating Index. *ASCE Journal of Management in Engineering*, 33(2).

Cantor, R., and Parker, F. (1994). The Credit Rating Industry. *FRBNY Quarterly Review*, Summer–Fall, pp. 1–26.

Clevenger, C., Ozbek, M. E., and Simpson, S. (2013). *Review of Sustainability Rating Systems Used for Infrastructure Projects*. Associated Schools of Construction, 49th ASC Annual International Conference Proceedings.

Construction Industry Institute (CII). (1994). Pre-Project Planning: Beginning a Project the Right Way. Austin, TX: Construction Industry Institute.

Construction Industry Institute (CII). (2008a). PDRI: Project Definition Rating Index for Buildings. Austin, TX: Construction Industry Institute.

Construction Industry Institute (CII). (2008b). PDRI: Project Definition Rating Index Industrial Projects. Austin, TX: Construction Industry Institute.

Construction Industry Institute (CII). (2010). PDRI: Project Definition Rating Index for Infrastructure Projects. Austin, TX: Construction Industry Institute.

Forsberg, A., and Malmborg, F. V. (2004). Tools for Environmental Assessment of the Built Environment. *Building Environment*, 39(2), pp. 223–228.

Krekeler, P., Nelson, D. A., Gritsavage, J. S., Kolb, E., and McVoy, G. R. (2010). Moving towards Sustainability: New York State Department of Transportation's GreenLITES Story. *Green Streets and Highways*, 2010, pp. 461–479.

Pantelias, A., and Roumboutsos, A. (2015). A Conceptual Framework for Transport Infrastructure PPP Project Credit Assessments. *Journal of Finance and Economics*, 3(6), pp. 105–111.

Roumboutsos, A., Pantelias A., and Sfakianakis, E. (2016). A Methodological Framework for the Credit Assessment of Transport Infrastructure Projects. In Roumboutsos A. (Ed.), *Public Private Partnerships in Transport: Trends and Theory*. Abingdon: Routledge. ISBN: 978-1-315-70872-0, pp. 320–338.

7 Conclusions and recommendations

Athena Roumboutsos, Hans Voordijk and Aristeidis Pantelias

7.1 The big picture

All infrastructure projects need to have a 'raison d'être'. As a minimum, they need to secure that their benefits to society will be greater than the cost of delivering the project and, in that respect, a CBA is usually the starting point in a long process that ultimately leads to their implementation. Transport projects are no different especially since the service they provide bears the characteristics of a 'public good', increasing social welfare and bolstering economic development. As governments find it increasingly hard to deliver transport infrastructure on their own, value for money (VfM) assessments seek to explore and justify the involvement of the private sector in what has traditionally been perceived as an area of public sector responsibility. Ultimately, the goal is to strike a multi-stakeholder win balance that broadly maintains an alignment among the interests of society, the government and the private sector, when involved.

But are these initial assessments enough? Can they safeguard and/or guarantee the attainment of project goals and uphold initial predictions or projections of benefits and returns?

Experience shows otherwise. Despite the rigour with which these assessments may be conducted, it is intuitively well known, even by stakeholders with limited experience in transport infrastructure project implementation (but also in other infrastructure sectors too), that existing decision-support tools are less than perfect.

As presented in the introduction of this book, an apparent lack of decision-support tools that would guide project stakeholders through the phases of its implementation (i.e. procurement, construction and operation) currently exists. Filling this assessment gap has been a key motivation for the development of this book and can be related to the mismatch that has been identified to exist between demand for infrastructure investment and supply of infrastructure finance.

Effectively, the fundamental task that this book set forth to achieve was the presentation of a novel indicator that describes the *ability of a transport infrastructure project to withstand, adjust and recover from changes within its structural elements with respect to its ability to deliver specific outcomes*. The rationale behind the task was simple: existing decision-support tools for transport infrastructure investment seem to lack the ability to assess the likelihood of a project achieving its target performance outcomes. This book aims to fill this identified gap through the development, elaboration and illustration of the TIRESI.

Getting there required a number of intermediate steps, comprising various analyses that have been elaborated upon in some of the previous chapters of this book. From the consideration of this work, numerous points of interests can be brought forward, many of which can in fact be considered themselves as intermediate value-adding decision-support tools. These points are enumerated and briefly summarised below.

1 The transport infrastructure delivery, operation and maintenance system

The book presents the conceptualisation of transport infrastructure delivery, operation and maintenance as a system of interrelated and interacting elements within an implementation context, that is the 'Business models for ENhancing funding and Enabling Financing for Infrastructure in Transport' (BENEFIT) Framework, which is also the title of this book (Chapter 1). The system comprises the following elements: the transport infrastructure business model; the funding scheme produced by the business model; the financing scheme that is meant to match the risk profile of the funding scheme; the governance arrangement regulating all project-related activities; the financial economic context reflecting (among other things) the productivity of the environment (country) the project is built and operated in and the institutional context, reflecting the strength and maturity of the institutions regulating and promoting project-related activities and investments in general. Finally, the transport infrastructure mode typology defines the business models that are applicable when the system is designed. During implementation, the influence of the transport mode is expressed by the system's availability and reliability.

2 The validation of the BENEFIT Framework as a useful and powerful explanatory tool for analysing, understanding and interpreting factors affecting project performance

The overview of findings from the mode-specific qualitative analysis (Chapter 2) not only contributed to the demonstration of how factors influencing project performance across modes may be grouped under and related to the elements of the BENEFIT Framework, but also determined the degree of representativeness of the sample of the BENEFIT project case dataset used in the analysis. From this point on:

- the implementation context of a project is described by its surrounding financial economic and institutional setting. A simple reference to country names is deemed inadequate as implementation settings within a country may change during the project's lifetime;

- the positioning of the infrastructure asset in the transport network and the resulting level of 'exclusivity' of its corresponding service(s), that is its market (monopoly) status form part of the business model setup;
- actors' competences (whether public or private) are assessed with respect to their ability to drive the business model and control the risks allocated to each of them.

Indisputably, the BENEFIT Framework adds a fresh perspective to qualitative analysis. Its value is demonstrated through the consideration and interpretation of findings from the comparison of the separate mode-specific analysis conducted.

3 The representation of the BENEFIT Framework elements through carefully selected and operationalised quantifiable indicators

Although in-depth understanding of many interactions of the BENEFIT Framework elements was achieved through the qualitative analysis per mode, limitations were still apparent, especially with respect to combining and reconciling different pieces of qualitative information. Overcoming these limitations by seeking to conduct a quantitative analysis was the next target, but this approach also carried the risk of dealing with an unmanageable number of identified factors. The answer on how to pool together qualitative information and how to deal with an increasing number of variables in order to enhance the explanatory value of the BENEFIT Framework is the introduction of the BENEFIT concept and its indicators (Chapter 3). Effectively, nine indicators numerically represent the six elements of the BENEFIT Framework and provide a quantifiable measure of their key attributes. Indicators are distinguished as exogenous (two), referring to the implementation context element; structural (three), concerning the business model and governance arrangement elements; policy (three), reflecting the funding and financing scheme elements; while a final indicator represents the impact of the transport mode typology on project implementation.

Through the introduction of indicators, new analysis possibilities become possible. The approach to be followed, however, needed to avoid pitfalls and mitigate weaknesses found in common numerical analysis methods. Effectively and in order to address these concerns, a multi-method analysis of the BENEFIT project case dataset was conducted employing quantitative and semi-quantitative methods. The pooling of respective findings demonstrated the advantages of a multi-method analysis compared to a single analysis approach, especially when addressing relatively small datasets.

Notably, by considering the indicator-based analysis findings and the resulting observations on element interactions, it was confirmed that there is no single indicator that is dominant in driving the attainment of different outcomes but rather combinations of indicators. Additionally, there is no single combination of indicators that can secure the successful attainment of all project outcome targets simultaneously and, finally, the implementation context indicators, which are external to the project, are always important but not determining.

4 How indicators work

On a first instance, while indicators have been designed to reflect lower project cost and risk as their values progress from low to high (i.e. from 0 to 1), is there always a positive correlation between high indicator values and support in the attainment of project performance targets? In a partly surprising, partly reassuring way, this does not hold true for all indicators. More specifically, indicators describing the funding (remuneration and revenue) and the financing scheme of a project, depending on their values and in combination with other indicators, place limitations and/but, also, induce incentives to project performance (Chapter 4). Effectively, these indicators are closely related to policies adopted within the implementation of the project, such as the introduction of infrastructure pricing and user charges and the level and type of involvement of private (but also public) sector financiers in project delivery.

On a second instance, are there cut-off indicator values that can be related to a positive or negative impact on attaining project outcome targets? Indeed, specific indicator values can be used as benchmarks determining their positive, negative or no influence on project performance, either on their own or in combinations. These threshold values were determined through a combination of numerical calibration methods related to the attainment of outcome targets (Chapter 4).

5 Addressing each transport infrastructure mode

New findings generate new questions needing answers. While some commonalities may be observed among transport infrastructure modes, which combinations of indicators (including their value ranges) drive the attainment of project targets in each case? The answer comes heuristically by systematically combining, comparing and synthesising findings from multiple sources: existing literature, results from the qualitative analysis, results from the indicator-based multi-method analysis as well as the in-depth screening of project cases in the BENEFIT project case dataset, providing the drivers of resilience for each transport infrastructure mode (Chapter 4).

6 The Transport Infrastructure Resilience Indicator

By identifying mode-specific drivers of performance attainment, or, viewed from another side, drivers of project resilience, and their threshold values, the ground is set for the introduction of the TIRESI. The methodological underpinnings of the TIRESI are based on academic research in project and system resilience and, also, on industry-based credit assessment methodologies with respect to how ratings scales are constructed. Hence, the indicator is both well-founded scientifically and easy to perceive in terms of what its scope is (Chapter 4).

The TIRESI anticipates that decision makers have different needs and, thus, has three manifestations, namely static (S-TIRESI), dynamic (D-TIRESI) and overall (O-TIRESI). Each version conveys different information and has a different role to play as a decision-support tool. The quantification methodology of the TIRESI is based on the development of a rating system, which aims to classify/categorise projects based on their likelihood of attaining specific outcome targets (Chapter 4).

7 Examples

Having the TIRESI methodology in place, its usefulness, explanatory power and applicability are demonstrated through the elaboration of nine project case studies from various transport modes (Chapter 5).

Through the elaboration of these cases, a number of important observations are made:

- the importance of the procurement procedure followed as this is reflected in the contractual arrangement and the significance of the contracting authority as planner, procurer and monitoring agent;
- the significance of the implementation context (financial economic and institutional) in the attainment of project targets: a positive implementation context is supportive but not a pre-requisite for the attainment of project targets. In other words, there are positive investment opportunities in all countries irrespectively of their financial economic and institutional characteristics, albeit under specific conditions and
- the potential of policy indicators to enhance the likelihood of achievement of project targets. Policy indicators represent the prevailing (funding and financing) policies accumulating and balancing stakeholder interests.

8 Filling the project assessment gap

The TIRESI rating system fills a gap with respect to the range and coverage of existing decision-support tools available to those investing in the 'business' of transport infrastructure, as well as those investing in transport infrastructure 'as an asset class' (Chapter 6). The BENEFIT Framework and the TIRESI can serve in combination as a valuable decision-support tool for various stakeholders in different phases of the delivery and implementation of transport infrastructure projects. This goes full-circle to the basic premise that this book started off from: the treatment of the mismatch between demand for infrastructure investment and supply of infrastructure finance. The discussion of this decision-support potential together with relevant examples, form the final contribution of this book to its audience, both academic and practitioner based (Chapter 6).

The chapters of this book are linked to one another constituting stepping stones in the journey to understand, operationalise and apply the BENEFIT Framework and the TIRESI. This effort has ultimately culminated in the development of an online Transport Infrastructure Resilience Assessment Tool (TIRESIAS). At the same time, each chapter provides for a concise reading with conclusions and considerations that may be immediately acknowledged by the readers while reflecting on their respective interests.

The aim of this last conclusive chapter is to provide a final contribution to transport infrastructure project stakeholders. It summarises the key conclusions and contributions of this book per interest group and the way these may ultimately assist and hopefully enhance their decision-making process. Therefore, the remainder of this chapter concerns 'takeaways'.

Finally, a key takeaway is the contribution of this book in resolving the mismatch between demand for infrastructure investment and the supply of infrastructure finance. This is discussed, separately, in the last section of this chapter.

7.2 Takeaways

As described in Chapter 6, involvement in the delivery and operation of transport infrastructure may broadly be of interest to two stakeholder groups: (i) those involved in 'infrastructure as a business', and therefore including public sector agencies, construction firms, architectural and planning firms, asset operators, project sponsors, etc.; and (ii) those interested in transport infrastructure 'as an asset class', that is investors who are solely interested in the financial returns that can be produced by and captured from these assets. These two groups differ not only in their level of expertise, with respect to their understanding of the technical, economic and financial characteristics of these assets, but also in their short- and/or long-terms interests. Notably, knowledge and interest asymmetries exist not only between, but also within these two groups of stakeholders. In addition to the above two groups, the work presented in this book may also be of interest to academic researchers. Some takeaways are also formulated for them.

7.2.1 Takeaways for researchers

A key concept streaming from the BENEFIT Framework has been the conceptualisation of transport infrastructure delivery, operation and maintenance as a system. The approach has allowed for new insights on and an enhanced understanding of transport infrastructure projects, which distances itself from overrating the influence of a single or a restricted number of factors on project performance, while capturing their complex interactions. It also allows for new methodological tools to be introduced in the study of projects and project management. Notably, systems thinking helps to identify and understand various interdependencies, interactions and feedback dynamics between the components of a defined system and their impacts on system performance (Locatelli *et al.*, 2014; Zhu and Mostafavi, 2014, 2017).

While improving understanding, systems-thinking inevitably imposes additional modelling complexity. In the case of the BENEFIT Framework, complexity was handled through the introduction of composite indicators, which reflect the key attributes of the identified system elements. Important contributions to

research were made in the process of constructing, validating and operational-
ising these indicators.

In terms of capturing the real-life dynamic nature of transport infrastructure
delivery and operation, this was addressed by representing and studying projects
through a series of 'snapshots'. By using 'snapshots', the evolution of project
conditions, characteristics and performance was captured at different points
in their lifetime through the observed variation of the corresponding indicator
values.

In addition, in this research, a multi-method analysis approach was adopted.
Multi-method analysis proved to be rigorous, comprehensive, and ultimately
valuable by building on the advantages of each individual method and compensating for its respective shortcomings. The approach, also known as 'triangulation',
is not new (Jick, 1979). In effect, mixing quantitative and qualitative research
methods is desirable, especially when relatively small datasets are concerned, to
validate whether the 'variation reflects that of the trait or the method' (Campbell
and Fiske, 1959 in Jick, 1979).

The TIRESI has been developed following a heuristic approach and bears
respective advantages and limitations. It addresses the issue of sample size and
the multiplicity of interactions between factors of the system and allows problem
solving in terms of a 'satisfying' solution, as it would be defined by Herbert A.
Simon (Augier and March, 2004). What is also important, however, is that this
approach allows for 'continuous learning'. The TIRESI's accuracy will continue
to improve as the number of cases analysed by it continue to increase.

In its current state of development, TIRESI is able to provide ratings for four
outcomes (cost and time to completion, traffic and revenue forecasts) in the case
of road, bridge and tunnel, and urban transit projects. It is also capable of
providing ratings for cost and time to completion for airport projects. Within the
current effort, rail project outcomes could not be rated as insufficient data was
available for analysis. All port project outcomes as well as airport project traffic
and revenue outcomes require an adjustment of the indicator composition in order
to represent the market interrelations of these transport infrastructure modes. As
a result, for these cases/modes, the TIRESI could also not provide meaningful
ratings.

Finally, the BENEFIT Framework as a description of the transport infrastructure
delivery and operation system is sufficiently generic and, as a system concept,
may thus be transferred to other infrastructure sectors (e.g. energy).

7.2.2 Takeaways for stakeholders in the 'transport infrastructure business'

7.2.2.1 Quantifying experience: the Transport Infrastructure Resilience Indicator and rating methodology

Considerable research has been conducted and professional experience exists with
respect to transport infrastructure project implementation. The effort presented

in this book has been to build on previously reported and well-known findings. It constructively combines these sources to produce quantifiable information for decision makers. Hence, the essential added value of the TIRESI is its capacity of transforming previously fragmented and discerned knowledge and experience into a useable and applicable tool.

Furthermore, the TIRESI and the BENEFIT Framework indicators have been developed based on project information that is readily available in the public domain. They do not require proprietary information or highly detailed data to produce results, which makes their implementation very straightforward. In its current formulation, the TIRESI is well positioned to:

- assist in building and testing various ex-ante project implementation scenarios providing support to:
  - public authorities to:
    - ○ better allocate risks or assess the influence that project structure decisions may have on the project's potential of reaching specific outcome targets under various implementation context conditions. Through this process, it may also allow for the identification of adverse factors and the specification of corresponding mitigation and other performance-enhancing actions, including modifications to funding and financing schemes;
    - ○ create improved and supportive project procurement processes;
    - ○ assess alternative implementation scenarios under (re) negotiations.
  - private parties to:
    - ○ consider and evaluate investment options;
    - ○ investigate and gauge the impact of their involvement on project outcomes;
    - ○ assess alternative scenarios under (re)negotiations.
- assist in monitoring project 'health' during implementation and operation;
- assist in estimating the impact of new financing and funding schemes on project outcomes due to the TIRESI's capability to consider both current as well as potential future funding and financing schemes;
- improve the creditworthiness of a project as it provides information with respect to the ability of a project to mitigate downside impacts through risk management rather than by placing the emphasis on its financing structure. In this context, the TIRESI can act complementarily to existing commercial credit ratings, as the combination of the two ratings (Credit and TIRESI) can provide a more comprehensive assessment of project resilience: managerial and financial.

Notably, the TIRESI guides decisions as to which factors need to be improved or changed in order to reach desirable project outcomes. During this process, the TIRESI will also indicate which project characteristics cannot be further improved thus minimising or eliminating the cost of ineffective actions and interventions. Ratings should also be used in combination with the identified, in each case, resilience drivers to support decisions on actionable project elements and factors. Obviously, certain project characteristics (and therefore the corresponding indicator values) are difficult or undesirable to change as time elapses in a project's lifetime. TIRESI ratings may then be useful in evaluating trade-offs between attaining different project target outcomes.

7.2.2.2 BENEFIT Framework structural indicators as project benchmarks

As previously mentioned, the key elements that describe the transport infrastructure and delivery system are as follows: the implementation context, the business model, the funding scheme, the financing scheme, the contractual governance conditions of implementation and, finally, the transport mode context. These elements are represented by nine composite indicators that have been developed and validated as presented in Chapter 3. These indicators, apart from being an essential part in the assessment of the TIRESI, may also serve as project benchmarks.

More specifically, structural indicators allow during all phases of project development the assessment of whether the project structure permits, and to what level, the exploitation of the project's full potential. The corresponding indicator threshold values are also important but should not be considered as absolute prerequisites. More specifically:

- The *Cost Saving Indicator* (CSI) illustrates a measure of a project's efficiency during construction and operation. In effect, it includes:
  - the ability to construct of the relevant private sector actor (level of civil works/technical difficulty; capability to construct based on the market position of the contractor with respect to construction or respective project delivery capability [example for rolling stock]; construction risk allocation as per contractual agreement; assessment of optimal construction risk allocation based solely on the capability to construct) and also the adoption of innovation and its successful application;
  - the ability of the relevant public or contracting authority to monitor/ control/plan and provide political support;
  - the ability to operate of the operator (private or public actor; life cycle planning verification; capability to operate based on the market position of the operator; operation risk allocation as per contractual agreement; assessment of optimal operational risk allocation based solely on the capability to operate).

- The *Revenue Support Indicator* (RSI) is also a composite indicator that may be considered a measure of the project's ability to generate revenues, and, also a measure of the project's efficiency in exploiting its potential sources of revenue. It includes the level of coopetition[1] of new (greenfield) and/or existing (brownfield) parts of the project, expressing the level of business development scope designed to attract demand (e.g. airports etc.); the level of project exclusivity with respect to its position in the transport network (e.g. metros, bridge and tunnel projects, ports, airports under certain conditions) and the level to which a transport network supports the project's exclusivity. The RSI also includes revenue sources attached to the project (traffic from new and brownfield operation, traffic from other transport infrastructure bundled in the project, as well as revenues related to non-transport services, all in relation to the capability to manage demand, demand risk allocation, assessment of demand risk allocation based on the capability to manage demand, quality of service).

- The *Governance Indicator* (GI), which contains factors that reflect many aspects of the relationship between the contracting authority and the contractor(s)/concessioner(s) refers to aspects of project governance such as early involvement of the contractor/concessionaire in the design and in the estimation of costs, procurement procedures, integration of design and construction, the incentives and disincentives regime, risk sharing, contract flexibility, and actions that enable the contracting authority to retain bargaining power during potential renegotiations.

7.2.2.3 Demystifying project resilience drivers

Resilience drivers leading to the attainment of specific project outcomes have been presented in the previous parts of this book with respect to their overall consideration (Chapter 3) and per infrastructure mode (Chapter 4). These drivers (effectively indicators or combinations of them) can be considered as takeaways on their own as they provide knowledge of their effects per transport infrastructure mode. However, there are a few additional relevant points that are of particular interest, which also place a number of best practices into context.

As often mentioned, in project management 'one size does not fit all'. Therefore:

- There is no single factor that can define the likelihood of achieving an outcome target but rather combinations of them. Hence, the implementation context and/or the creditworthiness of the project sponsors do not define the potential of a project to reach target outcomes on their own.
- There is no single combination of factors that can secure the successful attainment of all project outcome targets simultaneously. This suggests that trade-offs are common and should be expected during project implementation.
- Outcome targets are not achieved by the same combination of factors across all modes of transport. Therefore, each infrastructure mode requires respective expertise.

More specifically, each transport infrastructure mode is influenced differently by the implementation context and different factors contribute in each case to achieving the respective outcome targets. The difference lies primarily in the ability to fully endorse factors identified to support the achievement of project outcomes.

Nonetheless, some indicators were found to be supportive across all modes. These are the Institutional, Governance, Cost-Saving and Revenue Support Indicators. In this respect, and in order to increase the likelihood of achieving the envisaged project performance, governments' efforts are well placed to:

- strengthen the procurement (tendering and award) process;
- design contracts that include performance criteria and adequate, in each case, flexibility by:
  - including contractor selection criteria referring to respective competences;
  - enhancing the capabilities of contracting authorities to plan and procure projects as well as monitor their implementation. In this context, standardisation of procedures (both in project planning and procurement) and contracts may prove a useful tool for local authorities, who may lack respective competences. Also, with respect to PPPs, PPP Units as knowledge sharing and support institutions are important;
- appropriately share project risks, based on parties' ability to control them;
- properly integrate infrastructure in the transport network;
- continuously improve institutions on a national, regional and local level. Institutions are of extreme importance, especially with respect to political stability, control of corruption, accountability, regulation as well as other supportive policies.

Additionally, achieving performance becomes even more complicated when considering that macro-economic conditions, country competitiveness and productivity (captured by the Financial–Economic Indicator – FEI), while beyond the control of project decision makers (i.e. exogenous to the project) and significantly influential on project performance, are not a decisive factor in attaining project outcome targets. If sufficiently mature, the institutional context may balance out negative impacts stemming from lower values of the financial economic context. A similar impact, as that of the institutional context, was observed stemming from the Governance (GI) and Cost-Saving Indicators (CSIs). These observations have two major ramifications:

- Countries with sufficiently mature institutions are better positioned to withstand fluctuations in their financial economic context.
- Projects with a robust structure (governance, cost saving and, also, Revenue Support Indicators) are capable of succeeding even in poor implementation environments.

The contrary is also valid. That is to say, a poor project structure (low values of the governance, cost saving and, also, Revenue Support Indicators) may impede a project's ability to attain its target performance. Hence, it is the actionable indicators (reflecting elements endogenous to the project) that will ultimately determine the likelihood of positive performance.

Actionable indicators are categorised as structural and policy. While, structural indicators are mostly defined during the project's planning, tendering and award stage (project maturity, business model and contractual conditions/configuration), policy indicators (financing structure, project income (remuneration scheme) and project revenue streams) are actionable (under conditions) throughout the project life cycle and induce trade-offs with respect to the achievement of project outcomes.

Notably, the potential impact of introducing user charges for transport services and their level have been extensively studied with considerable knowledge and experience existing to this effect. At the same time, the financing scheme is often determined through the balancing of financiers' and project sponsors' risk perceptions and risk tolerance. Their combination often leads to costly financing. However, a key finding, confirming policy makers' anxiety, is the fact that low-cost financing is required to support the attainment of project objectives.

Values of the Financing Scheme Indicator (FSI) lower than 0.60 were found to be unsupportive of project targets. Considering that FSI = 1 represents a fully publicly financed project, there is considerable space for both public and private (co)financing of projects. However, if private financing comes at considerable cost, then in order to achieve values greater than 0.60, substantial public funds are required. In essence, this goes against current governments' efforts of introducing private financing for infrastructure, as it appears that they still need to provide a considerable part of the financing in order to deliver projects with a higher likelihood of attaining their performance targets. Additionally, in cases of public and private (co)financing solutions, a significant part of a project's revenue streams (direct and/or indirect) is required to repay private sector investments. This makes VfM assessments important and necessary to determine and indeed confirm the existence of benefits accruing to the public sector from these procurement and financing solutions.

In the context described in this section, it is not surprising that achieving transport infrastructure project performance was found to be rather independent of the underlying financing scheme (traditional public procurement or PPP). In other words, projects with FSI values within the entire range of [0.6, 1], thus ranging from privately (co)financed (PPP) to purely publicly financed, did not present any particular trend with respect to a higher likelihood of attaining project targets at either side of the range. Instead, the conditions of improved performance are mostly related to the actual project characteristics and the competences of the involved parties. In all cases, the competence of the public contracting authority and the appropriate sharing of responsibility (risks) among the involved parties, based on their level of control over risk, were found to be of paramount importance.

7.2.3 Takeaways for stakeholders investing in transport infrastructure 'as an asset class'

The BENEFIT Framework and the TIRESI ratings can be particularly useful to investors that are interested in the financial returns that can be generated by and captured from transport infrastructure projects.

First, policy indicators (i.e. funding and financing indicators) can be directly linked with investment decisions. Funding scheme indicators (i.e. revenue robustness and Remuneration Attractiveness Indicators) capture the nature and riskiness of identified revenue streams and remuneration methods. Their values paint the overall funding risk profile of the project under consideration that is fundamental in determining the corresponding investment risk. In effect, investors and financiers are concerned not only about the ability of a project to generate revenues, but also about the effectiveness of the method used to capture and direct them to various stakeholders. Revenues are important as they represent the operating cash flows based on which the project will cover operating costs, service debt obligations and generate equity returns. Based on these cash flow projections and their riskiness, financing schemes will be crafted by considering also the risk appetites of the different contributing financiers and investors. Moreover, the TIRESI also allows for the testing of new financing schemes that can support private sector contribution to transport infrastructure financing.

Second, TIRESI ratings offer complementary information to existing investment-supporting tools, such as project credit ratings. While credit ratings focus predominantly on the project's ability to honour its debt obligations, TIRESI ratings touch upon additional aspects of project performance thus also being able to cater to the interests of equity investors and not just debt investors. Ultimately, all investors are interested to determine whether their expected returns will be realised or not. By understanding whether a project will be likely to achieve its performance targets or whether it has specific vulnerabilities, they may form a view on the impact of these vulnerabilities on their returns and thus decide accordingly on the type and extent of their exposure to the project.

Third, investment returns from transport infrastructure projects are not uniform or consistent. While research is still on-going in this area (as well as in other sectors of infrastructure investment), it is clear that not all transport modes have similar risks and therefore the corresponding returns should be expected to differ. The TIRESI ratings explicitly differentiate between different modes as the methodology has been developed and calibrated accordingly (although its explanatory power does not cover all modes with respect to all four project outcomes considered due to data availability limitations). Nevertheless, the fact that TIRESI ratings are mode- and outcome-specific should enhance investor confidence as it demonstrates its specialised and bespoke nature as a decision-support tool. On the other hand, as was demonstrated in Chapter 5, under conditions, infrastructure modes for which the TIRESI ratings are not currently available, may also be assessed.

Finally, projects that reach their outcome targets usually enjoy the support and obtain high acceptance rates from all involved stakeholders. This is particularly true for transport infrastructure projects whose outcomes are scrutinised almost daily by their users and whose perceived success or failure is many times the subject of public discourse. Poorly performing projects, on the other hand, suffer from low public acceptance and have higher political risk in the sense that a public authority may feel under pressure to modify or terminate a project that is not delivering good outcomes in the eyes of the public, even if (or especially when) it is delivering its expected returns to its private investors. The TIRESI ratings, by capturing various aspects of project performance, can provide investors with a proxy assessment of the likelihood that their project of interest may run into difficulties, some of which may be deemed especially relevant to political or public acceptance risk. They can therefore be used as a useful ex-ante assessment and life-cycle monitoring tool for the types of investment risks that may affect the terms and conditions of an existing agreement and thus their ability to generate their expected returns.

7.3 Addressing the identified mismatch of investment demand and supply of financing for infrastructure projects

The BENEFIT Framework and the TIRESI ratings can be perceived to provide a new bridge to the assessment gap identified to exist within the universe of infrastructure investment decision-support tools. This is important in light of the basic premise of this book, which refers to the mismatch that has been found to exist between demand for infrastructure investment and supply of infrastructure finance. This mismatch was based on three key observations/reasons: stakeholders' lack of proper sector-related expertise; the over-reaching hand of financial regulation; and the uncertainty regarding the economic fundamentals supporting new and existing projects. The BENEFIT Framework and the TIRESI ratings could play a role in alleviating the severity of all these reasons behind the mismatch as explained below.

First, the BENEFIT Framework and the TIRESI ratings provide an explanatory framework for understanding and interpreting the impact of internal and external project characteristics on anticipated performance. Their development has gone through a complex and scientifically rigorous process. The use of the resulting TIRESI Assessment (TIRESIAS) tool is straightforward as it requires simple and easy to find inputs to calculate the values of the nine (9) indicators and produce project ratings. After all, the tool itself was developed based on project information found in the public domain, thus not relying on exclusive and potentially proprietary project information. Consequently, by using the framework and the tool, even stakeholders with limited previous exposure in the transport sector can obtain a first-hand understanding of areas of vulnerability or strength exhibited by their project of interest. Although using the TIRESIAS tool cannot be considered as formal training on transport infrastructure delivery, the information produced by the tool would help enhance stakeholders' understanding in terms of the likelihood of their project attaining its anticipated performance targets and support their corresponding investment decisions. In this context, this book can be considered to partially mitigate the first issue behind the mismatch, that is lack of stakeholder expertise and understanding of transport infrastructure projects.

Second, much of the financial regulation that has driven the restrictions in the provision of infrastructure finance has been based on poor project performance as reflected by its effect on financial investors and institutions. Some of this evidence has been corroborated by the downward revision of project credit ratings, while some has been anecdotal and/or kept confidential by investors due to its commercial sensitivity. Ultimately, however, regulators take stock of various market signals and craft their policies based on and in response to them. The TIRESI aspires to serve as one additional market signal that in combination with already existing ones can assist regulators in understanding the true perils that would endanger the successful delivery and implementation of transport infra-structure projects and, thus, inform their policies accordingly. Ultimately, when regulation becomes over-reaching, it has unintended consequences and adds government failures[2] to the market failures it tries to fix. In this respect, TIRESI ratings could be valuable as a source of information that can show how exposed financial institutions and individual investors contributing funds to transport infrastructure projects are with respect to the likelihood of these projects achieving (or not) their project outcome targets.

Finally, economic fundamentals should be at the forefront of the discussion behind the approval, delivery and implementation of all infrastructure projects. The uncertainty behind available funding streams that could support the financing of new projects, especially in the aftermath of the Global Financial Crisis, has become a fact of life for many public procuring agencies as well as private infrastructure investors. The BENEFIT Framework and TIRESI ratings explicitly take into consideration not only the nature and risk of the underlying revenue streams of a project of interest, but also the nature and risk of the remuneration method by which these streams can be captured and transformed into payments for the various project stakeholders. In doing so, it highlights relevant vulner-abilities and enables project stakeholders to investigate and/or identify additional sources of revenue that could support a project to strengthen its financial viability and, thus, make its delivery possible. In that respect, and although it does not alleviate the impact of global financial conditions on people's and counties' willingness and ability to pay for new infrastructure projects, the use of these tools enables a productive investigation of funding solutions that would be able to make a project viable and, thus, implementable.

7.4 Concluding remarks

In practical terms, TIRESI ratings reflect likelihoods and not an absolute certainty. Consequently, project stakeholders, whether public authority officials, con-struction project managers, sponsors and/or financiers should consider ratings as signals for taking relevant actions. A poor rating should be perceived as a warning sign reflecting the need for close project monitoring and effective risk manage-ment. A good rating should give confidence that a project is doing well although stakeholders should always remember that ratings are time-specific. A change in project implementation conditions and/or characteristics (whether internal or external) may change a rating and thus necessitate different future actions. Consequently, the TIRESI ratings are meant to be used as a guiding tool to be employed by all stakeholders throughout the project lifetime.

7.5 Notes

1   The term 'coopetition' is used to denote the nature of competition between the various parts (projects) of the transport network, where both 'cooperation' and 'competition' exists.
2   In the analysis of regulation, government failure (or non-market failure) is imperfection in government performance (Orbach, 2013).

7.6 References

Augier, M., and March, J.G. (2004). Models of Man, Essays in Memory of Herbert A. Simon. Cambridge, MA and London: The MIT Press, ISBN 0-262-01208-1.

Jick, T.D. (1979). Mixing Qualitative and Quantitative Methods: Triangulation in Action. *Administrative Science Quarterly, Qualitative Methodology*, 24(4), pp. 602–611.

Locatelli, G., Mancini, M., and Romano, E. (2014). Systems Engineering to Improve the Governance in Complex Project Environments. *International Journal of Project Management*, 32, pp. 1395–1410.

Orbach, B. (2013). What Is Government Failure. *Yale Journal on Regulation Online*, 30, pp. 44–56.

Zhu, J., and Mostafavi, A. (2014). A System-of-Systems Framework for Performance Assessment in Complex Construction Projects. *Organisation, Technology and management in Construction: An International Journal*, 6, pp. 1083–1093.

Zhu, J., and Mostafavi, A. (2017). Discovering Complexity and Emergent Properties in Project Systems: A New Approach to Understanding Project Performance. *International Journal of Project Management*, 35(1), pp. 1–12.

Appendix
BENEFIT project cases dataset

The dataset assembled for the purpose of the analysis of the case studies includes 86 cases, of which 55 are PPPs and 31 public projects.

The dataset has been created based on initially collected information from COST Action TU1001 (Roumboutsos *et al.*, 2013, 2014) and the Omega Centre Megaproject (see www.omegacentre.bartlett.ucl.ac.uk/publications/omega-case-studies). The case studies were updated to meet the needs of the BENEFIT project, while new cases were added (see BENEFIT Wiki and e-Book). Information was collected through surveys, interviews and desk research carried out by the consortium partners. Table A1 below lists all cases and their basic characteristics, namely the country of implementation, the selected delivery mode, the primary and other transport modes served by the infrastructure as well as the existence of greenfield and/or brownfield sections for private (co)financed projects. Table A2 provides the same information for projects financed totally by the public sector.

Notably, against expectations, the collection of information with respect to transport infrastructure projects publically financed and delivered through traditional procurement proved very cumbersome. Key reasons to this shortcoming were:

1   Infrastructure projects delivered through traditional procurement are seldom turn-key. They usually include multiple contracts, which are not registered in combination. Therefore, it is very difficult to identify information on all contracts related to the same infrastructure section.
2   Publically procured infrastructure projects are usually procured in small over-lapping sections making it difficult to define the 'project' and its outcomes.

Notably, with respect to the frequently mentioned issue of information sharing, it was found that it is not only related to operational confidentiality and/or international business information disclosure ethics. It is also a matter of systematic recording of data concerning not only PPP (or generally privately (co)financed) projects, but also projects financed by the public sector, for which even less information is available. The issue is rather related to lack of systematic data archiving and registering rather than information disclosure issues. The ramification is the inability to capitalise on existing experience and exploit relevant data.

Table A.1 List of the private (co)financed projects of the BENEFIT project case dataset

#	Project title	Country	Other mode	Field
1	Attiki Odos (Athens Ring Road)	Greece	Rail	Greenfield
2	Brebemi	Italy		Greenfield
3	Horgos – Pozega	Serbia		Both
4	Ionia Odos Motorway	Greece		Both
5	Central Greece (E65) Motorway	Greece		Both
6	BNRR (M6 Toll)	UK		Greenfield
7	M80 Haggs	UK		Both
8	A19 Dishforth To Tyne Tunnel	UK		Brownfield
9	A22 – Algarve	Portugal		Both
10	Radial 2 Toll Motorway	Spain		Greenfield
11	Eje Aeropuerto (M-12). Airport Axis Toll Motorway	Spain		Greenfield
12	M-45	Spain		Greenfield
13	A2 Motorway Poland	Poland		Greenfield
14	Istrian Y	Croatia		Both
15	A23 - Beira Interior	Portugal		Both
16	E39 Orkdalsvegen Public Road	Norway		Both
17	Elefsina Korinthos Patra Pyrgos Tsakona Motorway	Greece		Both
18	Via-Invest Zaventem	Belgium		Both
19	E18 Grimstad - Kristiansand	Norway		Both
20	M-25 Motorway London Orbital	UK		Brownfield
21	Moreas Motorway	Greece		Both
22	C-16 Terrassa-Manresa Toll Motorway	Spain		Greenfield
23	E4 Helsinki-Lahti	Finland		Both
24	E18 Muurla-Lohja	Finland		Greenfield
25	Rion-Antirion Bridge	Greece	Road	Greenfield
26	Lusoponte - Vasco Da Gama Bridge	Portugal	Road	Both
27	Coen Tunnel	The Netherlands	Road	Both
28	Herrentunnel Lüebeck	Germany	Road	Brownfield
29	Millau Viaduct	France	Road	Greenfield
30	The Oresund Link	Sweden – Denmark	Road Rail	Greenfield
31	Piraeus Container Terminal	Greece		Both
32	Port of Sines Terminal XXI	Portugal	Road Rail	Greenfield
33	Port of Leixoes	Portugal	Road Rail	Both
34	Deurganckdoksluis-Deurganckdock Lock	Belgium		Greenfield
35	Venice Offshore–Onshore Terminal	Italy		Both
36	Larnaka Port and Marina re-development	Cyprus	Marina and Real estate	Both
37	Valencia Cruise Terminal	Spain		Greenfield
38	Terminal Muelle Costa at Port of Barcelona	Spain	SSS, freight and passenger terminal	Greenfield

continued . . .

#	Project title	Country	Other mode	Field
39	Barcelona Europe South Terminal	Spain		Greenfield
40	Adriatic Gateway Container Terminal	Croatia		Both
41	Athens International Airport 'Eleftherios Venizelos'	Greece	Road Freight terminal	Greenfield
42	Larnaca and Paphos International Airports	Cyprus		Both
43	Fertagus Train	Portugal	Road Terminals Bus	Both
44	Liefkenshoekspoor-verbinding -Liefkenshoek Rail Link	Belgium		Greenfield
45	Metrolink LRT, Manchester	UK	Metro	Brownfield
46	Reims Tramway	France	Bus	Both
47	Caen-TVR	France		Both
48	Brabo 1	Belgium	Cycle lanes	Both
49	MST - Metro Sul do Tejo	Portugal		Greenfield
50	Metro de Malaga	Spain		Greenfield
51	Metro do Porto S.A.	Portugal	Funicular system	Both
52	Velo'V	France		Greenfield
53	SERVICI	Spain		Greenfield
54	Quadrante Europa Terminal Gate	Italy		Greenfield
55	Central Public Transport Depot of the City of Pilsen	Czech Republic		Brownfield

Table A.2 List of the projects financed by the public sector of the BENEFIT project case dataset

#	Project title	Country	Other mode	Field
1	Combiplan Nijverdal	The Netherlands	Rail	Brownfield
2	A5 Maribor - Pince Motorway	Slovenia		Both
3	Koper – Izola Expressway	Slovenia		Both
4	Motorway E-75, Section Horgos – Novi Sad (2nd Phase)	Serbia		Both
5	Belgrade By-Pass Project, Section A: Batajnica-Dobanovci	Serbia		Greenfield
6	Motorway E-75, Section Donji Neradovac – Srpska Kuca	Serbia		Brownfield
7	Estradas de Portugal	Portugal		Both
8	Bundesautobahn 20	Germany		Greenfield
9	Berlin Tiergarten Tunnel	Germany	Road Rail Terminals	Brownfield

continued . . .

Table A.2 Continued

#	Project title	Country	Other mode	Field
10	Sodra Lanken (The Southern Link)	Sweden	Road	Both
11	Blanka Tunnel Complex	Czech Republic	Road Tramway	Greenfield
12	OW-Plan Oostende-Integrated Coastal & Maritime Plan for Oostende	Belgium		Both
13	Port of Agaete	Spain	(concessioned operation)	Brownfield
14	Modlin Regional Airport	Poland	Rail Freight terminal	Brownfield
15	Berlin Brandenburg Airport (BER)	Germany	Road Rail	Both
16	Sa Carneiro Airport Expansion	Portugal	Road Terminals Metro station	Brownfield
17	Gardermobanen (Airport Exprestrain)	Norway	Tunnel Station	Both
18	Tram-Train 'Kombiloesung' Karlsruhe	Germany	Road Metro	Both
19	NBS Köln-Rhein/Main	Germany		Greenfield
20	TGV Mediterranean	France	Terminals	Both
21	HSL-Zuid	The Netherlands		Greenfield
22	MXP T2-Railink-up	Italy		Both
23	Tram T4 (Line 4 of Lyon Tramway)	France		Both
24	Athens Tramway	Greece		Greenfield
25	Randstadrail	The Netherlands	Road Terminals	Both
26	Warsaw's Metro II-nd Line	Poland		Greenfield
27	London Underground Jubilee Line Extension (JLE)	UK		Brownfield
28	Meteor	France		Both
29	Attiko Metro (Athens Metro Base Project)	Greece		Greenfield
30	Beneluxlijn	The Netherlands		Greenfield
31	The Hague New Central Train Station	The Netherlands	Rail	Brownfield

LRT = light rail transit.

Appendix references

Roumboutsos, A. (2015). Business Models for Enhancing Funding and Enabling Financing for Infrastructure in Transport: PPP and Public Transport Infrastructure Financing Case Studies. Horizon 2020 European Commission. Department of Shipping, Trade and Transport, University of the Aegean, Greece, ISBN 978-618-82078-1-3. www.benefit4 transport.eu

Roumboutsos, A., Farrell S., and Verhoest, K., (2014). COST Action TU1001 – Public Private Partnerships in Transport: Trends & Theory: 2014 Discussion Series: Country Profiles and Case Studies, ISBN 978-88-6922-009-8

Roumboutsos, A., Farrell, S., Liyanage, C. L., and Macário, R. (2013). COST Action TU1001 Public Private Partnerships in Transport: Trends & Theory P3T3, 2013 Discussion Papers Part II Case Studies, ISBN 978-88-97781-61-5.

Index